Statistics for
People Who *(Think They)*
Hate Statistics

4

EDITION

This book is dedicated with love and admiration to Sara, Micah, and Ted—simply the best—and to all the Sharks.

And, for Pepper

1994–2009
Outside of a dog, a book is man's best friend.
Inside of a dog, it's too dark to read.

—*Groucho Marx*

Statistics for
People Who (Think They)
Hate Statistics

4
EDITION

Neil J. Salkind
University of Kansas

Los Angeles | London | New Delhi
Singapore | Washington DC

This book includes screenshots of Microsoft Excel 2010 to illustrate the methods and procedures described in the book. Microsoft Excel is a product of the Microsoft Corporation.

For information:

SAGE Publications, Inc.
2455 Teller Road
Thousand Oaks, California 91320
E-mail: order@sagepub.com

SAGE Publications Ltd.
1 Oliver's Yard
55 City Road
London EC1Y 1SP
United Kingdom

SAGE Publications India Pvt. Ltd.
B 1/I 1 Mohan Cooperative Industrial Area
Mathura Road, New Delhi 110 044
India

SAGE Publications Asia-Pacific Pte. Ltd.
33 Pekin Street #02-01
Far East Square
Singapore 048763

Printed in the United States of America

Library of Congress Cataloging-in-Publication Data

Salkind, Neil J.
Statistics for people who (think they) hate statistics / Neil J. Salkind. – 4th ed.
    p. cm.
Includes bibliographical references and index.
ISBN 978-1-4129-7959-7 (pbk.)
ISBN 978-1-4129-7960-3 (paper w/SPSS CD)
    1. Statistics. I. Title.

HA29.S2365 2011
519.5–dc22   2010029241

This book is printed on acid-free paper.

10 11 12 13 14 10 9 8 7 6 5 4 3 2 1

Acquisitions Editor:	Vicki Knight
Associate Editor:	Lauren Habib
Production Editor:	Libby Larson
Copy Editor:	Liann Lech
Typesetter:	C&M Digitals (P) Ltd.
Proofreader:	Charlotte J. Waisner
Indexer:	Diggs Publication Services
Cover Designer:	Candice Harman
Marketing Manager:	Stephanie Adams
Permissions Editor:	Adele Hutchinson

BRIEF CONTENTS

PART IV

PART V

DETAILED CONTENTS

PART III

PART IV

PART V

A NOTE TO THE STUDENT: WHY I WROTE THIS BOOK

This is a new edition and a new time in my teaching career and I welcome you to what I hope will be, in all ways, a good learning experience.

What many students of introductory statistics (be they new to the subject or just reviewing the material) have in common (at least at the beginning of their studies) is a relatively high level of anxiety, the origin of which is, more often than not, what they've *heard* from their fellow students. Often, a small part of what they have heard is true—learning statistics takes an investment of time and effort (and there's the occasional monster for a teacher). But most of what they've heard (and where most of the anxiety comes from)—that statistics is unbearably difficult and confusing—is just not true. Thousands of fear-struck students have succeeded where they thought they would fail. They did it by taking one thing at a time, pacing themselves, seeing illustrations of basic principles as they are applied to real-life settings, and even having some fun along the way. That's what I tried to do in writing the first three editions of *Statistics for People Who (Think They) Hate Statistics,* and I tried even harder in completing this revision.

After a great deal of trial and error, and some successful and many unsuccessful attempts, I have learned to teach statistics in a way that I (and many of my students) think is unintimidating and informative. I have tried my absolute best to incorporate all of that experience into this book.

What you will learn from this book is the information you need to understand what the field and study of basic statistics is all about. You'll learn about the fundamental ideas and the most commonly used techniques to organize and make sense out of data. There's very little theory (but some), and there are few mathematical proofs or discussion of the rationale for certain mathematical routines.

Why isn't this theory stuff and more in *Statistics for People Who (Think They) Hate Statistics?* Simple. Right now, you don't need it. It's not that I don't think it is important. Rather, at this point and time in your studies, I want to offer you material at a level I think you can understand and learn with some reasonable amount of effort, while at the same time not be scared off from taking additional courses in the future. I (and your professor) want you to succeed.

So, if you are looking for a detailed unraveling of the derivation of the analysis of variance *F* ratio, go find another good book from SAGE (I'll be glad to refer you to one). But if you want to learn why and how statistics can work for you, you're in the right place. This book will help you understand the material you read in journal articles, explain what the results of many statistical analyses mean, and teach you how to perform basic statistical tasks.

And, if you want to talk about any aspect of teaching or learning statistics, feel free to contact me. You can do this through my e-mail address at school (njs@ku.edu). Good luck, and let me know how I can improve this book to even better meet the needs of the beginning statistics student.

AND A (LITTLE) NOTE TO THE INSTRUCTOR

This is the first time I have included a note to the instructor throughout all the editions of this book, and I would like to share two things.

First, I applaud your efforts at teaching these materials. Although they may be easier for some students, most find the material very challenging. Your patience and hard work are appreciated by all, and if there is anything I can do to help, please send me a note.

Second, *Statistics for People Who (Think They) Hate Statistics* is not meant to be a dumbed-down book similar to others you may have seen. Nor is the title meant to convey anything other than the fact that many students new to the subject are actually very anxious about what's to come. This is not an academic version of a book for dummies or anything of its kind. I have made every effort to address students with the respect they deserve, to not patronize them, and to ensure that the material is approachable. How well I did in these regards is up to you, but I want to convey my very clear intent and feeling that this book contains the information needed in an introductory course, and even though there is some humor involved in my approach, nothing about the intent is anything other than serious. Thank you.

ACKNOWLEDGMENTS

Everybody, and I mean everybody (including Steve in shipping and Sharon in contracts), at SAGE deserves a great deal of thanks for providing me with the support, guidance, and professionalism that takes only an idea (way back before the first edition) and makes it into a book like the one you are now reading—and then makes it successful.

However, there are some people who have to be thanked individually for their special care and hard work. Vicki Knight, senior editor, Research Methods and Statistics, has shepherded this edition, being always available to discuss new ideas and seeing to it that everything got done on time and done well. Lisa Cuevas Shaw, the previous editor, helped this book get to this point, and to her, I am forever grateful. Others who deserve a special note are Stephanie Adams, marketing manager; Lauren Habib, associate editor; Veronica Stapleton, production editor; and Libby Larson, production editor. Special, special thanks goes to Liann Lech for her sharp eye and sound copy editing, which make this material read as well as it does.

SAGE would like to thank the following peer reviewers for their editorial insight and guidance:

E. Helen Berry
Utah State University

Katherine M. Brown
University of New Haven

Lynnda J. Emery
Eastern Kentucky University

Elizabeth M. Flow-Delwiche
Community College of Baltimore County

Jennifer Pontius
University of Vermont

Amy Y. Spurlock
Troy University

Beverly J. Volicer
University of Massachusetts, Lowell

AND NOW, ABOUT THE FOURTH EDITION ...

What you read above about this book reflects my thoughts about why I wrote this book in the first place. But it tells you little about this fourth edition.

Any book is always a work in progress, and this latest edition of *Statistics for People Who (Think They) Hate Statistics* is no exception. Over the past 9 years or so, many people have told me how helpful this book is, and others have told me how they would like it to change and why. In revising this book, I am trying to meet the needs of all audiences. Some things remain the same, and some have indeed changed.

There are always new things worth consideration and different ways to present old themes and ideas. Here's a list of what you'll find that's new in the fourth edition of *Statistics for People Who (Think They) Hate Statistics*.

- It seems like everyone always wants more exercises for practice, and we hope that the new ones found at the end of each chapter do not disappoint. They vary in their level of application and (I hope) interest. These exercises use data sets that are available at www.sagepub.com/salkind4e and at the author's website at http://www.onlinefilefolder.com. Look for the Files tab. You can download them as needed. More about the files later.

- These data sets continue to come in two flavors—SPSS (that popular statistical analysis program) and Excel (the spreadsheet that many people use for data analysis). These data sets are available in Appendix C as well as here…

  1. online at http://www.sagepub.com/salkind4e, and

  2. at www.onlinefilefolder.com. The username is *ancillaries* and the password is *files*. Locate the files you want in the Excel or SPSS folder and then just right-click your mouse and select download.

The version of Excel for which these data sets were developed is Excel 2010 and SPSS version 19, but these files will work with earlier versions of both applications as well.

- A new appendix (Appendix E) specifically for readers who need to refresh their memory on basic arithmetic operations. This is short (only a few pages with some exercises) and is a good review.

- A new chapter! Lots of users wanted a chapter on the one-sample Z test, and this should have been included in an earlier edition but was not. It's added here (as Chapter 10) and follows the same format and pattern as the other chapters. We hope it works as a further introduction to inferential statistics.

- The fourth edition features SPSS 19, the latest version that SPSS offers. For the most part, you can use a version of SPSS that is as early as 11 to do most of the work, and even these earlier versions can read the data files created with the later versions. If the reader needs help on SPSS, we offer a mini course in Appendix A.

Any typos and such that appear in this edition of the book are entirely my fault, and I apologize to the professors and students who are inconvenienced by their appearance. And I so appreciate any letters, calls, and e-mails pointing out these errors. You can see all these errors at www.statisticsforpeople.com, and I welcome any additions. We have all made every effort in this edition to correct any previous errors and hope we did a reasonably good job. Let me hear from you with suggestions, criticisms, nice notes, and so on. Good luck.

Neil J. Salkind
University of Kansas
njs@ku.edu

ABOUT THE AUTHOR

Neil J. Salkind received his PhD from the University of Maryland in Human Development, and after teaching for 35 years at the University of Kansas, he remains as a Professor Emeritus in the Department of Psychology and Research in Education, where he continues to collaborate with colleagues and work with students. His early interests were in the area of children's cognitive development, and after research in the areas of cognitive style and (what was then known as) hyperactivity, he was a postdoctoral fellow at the University of North Carolina's Bush Center for Child and Family Policy. His work then changed direction and the focus was on child and family policy, specifically the impact of alternative forms of public support on various child and family outcomes. He has delivered more than 150 professional papers and presentations; written more than 100 trade and textbooks; and is the author of *Statistics for People Who (Think They) Hate Statistics* (SAGE), *Theories of Human Development* (SAGE), and *Exploring Research* (Prentice Hall). He has edited several encyclopedias, including the *Encyclopedia of Human Development,* the *Encyclopedia of Measurement and Statistics,* and the recently published *Encyclopedia of Research Design.* He was editor of *Child Development Abstracts and Bibliography* for 13 years and lives in Lawrence, Kansas, where he likes to read, swim with the River City Sharks, bake brownies (see the Excel version of *Statistics for People . . .* for the recipe at http://www.statisticsforpeople.com), and poke around old Volvos and old houses.

PART I

Yippee! I'm in Statistics

N ot much to shout about, you might say? Let me take a minute and show you how some very accomplished scientists use this widely used set of tools we call statistics.

• Michelle Lampl is a pediatrician and an anthropologist at Emory University. She was having coffee with a friend, who commented on how quickly her young infant was growing. In fact, the new mother spoke as if her son was "growing like a weed." Being a curious scientist (as all scientists should be), Dr. Lampl thought she might actually examine how rapid this child's growth, and others, is during infancy. She proceeded to measure a group of children's growth on a daily basis and found, much to her surprise, that some infants grew as much as one inch overnight! Some growth spurt.

Want to know more? Why not read the original work? You can find more about this in Lampl, M., Veldhuis, J. D., & Johnson, M. L. (1992). Saltation and stasis: A model of human growth. *Science, 258,* 801–803.

• Sue Kemper is a professor of psychology at the University of Kansas and has been working on the most interesting of projects. She and several other researchers studied a group of nuns and examined how their early experiences, activities, personality characteristics, and other information relates to their health during their late adult years. Most notably, this diverse group of scientists (including psychologists, linguists, neurologists, and others) wanted to know how well all this information predicts the occurrence of Alzheimer's disease. Among other really, really interesting findings, the group reports that the complexity of the nuns' writing during their early 20s is related to the nuns' risk for Alzheimer's 50, 60, and 70 years later.

Want to know more? Why not read the original work? You can find more about this in Snowdon, D. A., Kemper, S. J., Mortimer, J. A., Greiner, L. H., Wekstein, D. R., & Markesbery, W. R. (1996). Linguistic ability in early life and cognitive function and Alzheimer's disease in late life: Findings from the nun study. *Journal of the American Medical Association, 275,* 528–532.

• Aletha Huston is a researcher and teacher at the University of Texas in Austin and has devoted a great deal of her professional work to understanding what effects television watching has on young children's psychological development. Among other things, she and her late husband John C. Wright specifically investigated the impact that the amount of educational television programs watched during the early preschool years might have on outcomes in the later school years. They found convincing evidence that children who watch educational programs such as *Mr. Rogers* and *Sesame Street* do better in school than those who do not.

Want to know more? Why not read the original work? You can find more about this in Collins, P. A., Wright, J. C., Anderson, D. R., Huston, A. C., Schmitt, K., & McElroy, E. (1997). *Effects of early childhood media use on adolescent achievement.* Paper presented at the biennial meeting of the Society for Research in Child Development, Albuquerque, NM.

All of these researchers had a specific question they found interesting and used their intuition, curiosity, and excellent training to answer it. As part of their investigations, they used this set of tools we call statistics to make sense out of all the information they collected. Without these tools, all of this information would have been just a collection of unrelated outcomes. The outcomes would be nothing that Lampl could have used to reach a conclusion about children's growth, or Kemper could have used to better understand Alzheimer's disease, or Huston and Wright could have used to better understand the impact of watching television on young children's achievement and social development.

Statistics—the science of organizing and analyzing information to make the information more easily understood—made the task doable. The reason that any of the results from such studies are useful is that we can use statistics to make sense out of them. And that's exactly the goal of this book—to provide you with an understanding of these basic tools and how they are used—and, of course, how to use them.

In this first part of *Statistics for People Who (Think They) Hate Statistics,* you will be introduced to what the study of statistics is about and why it's well worth your efforts to master the basics—the important terminology and ideas that are central to the field. It's all in preparation for the rest of the book.

1 Statistics or Sadistics?

It's Up to You

Difficulty Scale ☺☺☺☺☺ (really easy)

WHAT YOU'LL LEARN ABOUT IN THIS CHAPTER

✦ What statistics is all about

✦ Why you should take statistics

✦ How to succeed in this course

WHY STATISTICS?

You've heard it all before, right? "Statistics is difficult," "The math involved is impossible," "I don't know how to use a computer," "What do I need this stuff for?" "What do I do next?" and the famous cry of the introductory statistics student, "I don't get it!"

Well, relax. Students who study introductory statistics find themselves, at one time or another, thinking about at least one of the above, if not actually sharing it with another student, their spouse, a colleague, or a friend.

And all kidding aside, there are some statistics courses that can easily be described as sadistics. That's because the books are repetitiously boring, and the authors have no imagination.

That's not the case for you. The fact that you or your instructor has selected *Statistics for People Who (Think They) Hate Statistics* shows that you're ready to take the right approach: one that is unintimidating, informative, and applied (and even a little fun) and that tries to teach you what you need to know about using statistics as the valuable tool that it is.

5

If you're using this book in a class, it also means that your instructor is clearly on your side—he or she knows that statistics can be intimidating but has taken steps to see that it is not intimidating for you. As a matter of fact, we'll bet there's a good chance (as hard as it may be to believe) that you'll be enjoying this class in just a few short weeks.

A FIVE-MINUTE HISTORY OF STATISTICS

Before you read any further, it would be useful to have some historical perspective about this topic called statistics. After all, almost every undergraduate in the social, behavioral, and biological sciences and every graduate student in education, nursing, psychology, social welfare and social services, and anthropology (you get the picture) is required to take this course. Wouldn't it be nice to have some idea from whence the topic it covers came? Of course it would.

Way, way back, as soon as humans realized that counting was a good idea (as in "How many of these do you need to trade for one of those?"), collecting information also became a useful skill. If counting counted, then one would know how many times the sun would rise in one season, how much food was needed to last the winter, and what amount of resources belonged to whom.

That was just the beginning. Once numbers became part of language, it seemed like the next step was to attach these numbers to outcomes. That started in earnest during the 17th century, when the first set of data pertaining to populations was collected. From that point on, scientists (mostly mathematicians, but then physical and biological scientists) needed to develop specific tools to answer specific questions. For example, Francis Galton (a cousin of Charles Darwin, by the way), who lived from 1822 to 1911, was very interested in the nature of human intelligence. To explore one of his primary questions regarding the similarity of intelligence among family members, he used a specific statistical tool called the correlation coefficient (first developed by mathematicians), and then he popularized its use in the behavioral and social sciences. You'll learn all about this tool in Chapter 5.

In fact, most of the basic statistical procedures that you will learn about were first developed and used in the fields of agriculture, astronomy, and even politics. Their application to human behavior came much later.

The past 100 years have seen great strides in the invention of new ways to use old ideas. The simplest test for examining the differences between the averages of two groups was first advanced during the early 20th century. Techniques that build on this idea were offered decades later and have been greatly refined. And the introduction of personal computers and such programs as SPSS®, an IBM company (see Appendix A), has opened up the use of sophisticated techniques to anyone who wants to explore these fascinating topics.

The introduction of these powerful personal computers has been both good and bad. It's good because most statistical analyses no longer require access to a huge and expensive mainframe computer. Instead, a simple personal computer costing less than $1,000 can do 95% of what 95% of the people need. On the other hand, less than adequately educated students (such as your fellow students who passed on taking this course!) will take any old data they have and think that by running them through some sophisticated SPSS analysis, they will have reliable, trustworthy, and meaningful outcomes—not true. What your professor would say is, "Garbage in, garbage out"—if you don't start with reliable and trustworthy data, what you'll have after your data are analyzed are unreliable and untrustworthy results.

Today, statisticians in all different areas from criminal justice to geophysics to psychology find themselves using basically the same techniques to answer different questions. There are, of course, important differences in how data are collected, but for the most part, the analyses (the plural of analysis) that are done following the collection of data (the plural of datum) tend to be very similar even if called something different. The moral here? This class will provide you with the tools to understand how statistics are used in almost any discipline. Pretty neat, and all for just three or four credits.

If you want to learn more about the history of statistics and see a historical time line, a great place to start is a Saint Anselm's College Internet site located at http://www.anselm.edu/homepage/jpitocch/biostatshist .html and http://www.stat.ucla.edu/history (at the University of California at Los Angeles). Tons of good stuff at both places.

STATISTICS: WHAT IT IS (AND ISN'T)

Statistics for People Who (Think They) Hate Statistics is a book about basic statistics and how to apply them to a variety of different situations, including the analysis and understanding of information.

In the most general sense, **statistics** describes a set of tools and techniques that is used for describing, organizing, and interpreting information or data. Those data might be the scores on a test taken by students participating in a special math curriculum, the speed with which problems are solved, the number of patient complaints when using one type of drug rather than another, the number of errors in each inning of a World Series game, or the average price of a dinner in an upscale restaurant in Sante Fe.

In all of these examples, and the million more we could think of, data are collected, organized, summarized, and then interpreted. In this book, you'll learn about collecting, organizing, and summarizing data as part of descriptive statistics. And then you'll learn about interpreting data when you learn about the usefulness of inferential statistics.

What Are Descriptive Statistics?

Descriptive statistics are used to organize and describe the characteristics of a collection of data. The collection is sometimes called a **data set** or just **data**.

For example, the following list shows you the names of 22 college students, their major areas of study, and their ages. If you needed to describe what the most popular college major is, you could use a descriptive statistic that summarizes their choice (called the mode). In this case, the most common major is psychology. And if you wanted to know the average age, you could easily compute another descriptive statistic that identifies this variable (that one's called the mean). Both of these simple descriptive statistics are used to describe data. They do a fine job allowing us to represent the characteristics of a large collection of data such as the 22 cases in our example.

Name	Major	Age	Name	Major	Age
Richard	Education	19	Elizabeth	English	21
Sara	Psychology	18	Bill	Psychology	22
Andrea	Education	19	Hadley	Psychology	23
Steven	Psychology	21	Buffy	Education	21
Jordan	Education	20	Chip	Education	19
Pam	Education	24	Homer	Psychology	18
Michael	Psychology	21	Margaret	English	22
Liz	Psychology	19	Courtney	Psychology	24
Nicole	Chemistry	19	Leonard	Psychology	21
Mike	Nursing	20	Jeffrey	Chemistry	18
Kent	History	18	Emily	Spanish	19

So watch how simple this is. To find the most frequently selected major, just find the one that occurs most often. And to find the average age, just add up all the age values and divide by 22. You're right—the most often occurring major is psychology (9 times) and the average age is 20.3. Look, Ma! No hands—you're a statistician.

What Are Inferential Statistics?

Inferential statistics are often (but not always) the next step after you have collected and summarized data. **Inferential statistics** are used to make inferences from a smaller group of data (such as our group of 22 students) to a possibly larger one (such as all the undergraduate students in the College of Arts and Sciences).

This smaller group of data is often called a sample, which is a portion, or a subset, of a population. For example, all the fifth graders in Newark, NJ, would be a population (it's all the occurrences with certain characteristics—being in fifth grade and living in Newark), whereas a selection of 150 of them would be a sample.

Let's look at another example. Your marketing agency asks you (a newly hired researcher) to determine which of several different names is most appealing for a new brand of potato chip. Will it be Chipsters? FunChips? Crunchies? As a statistics pro (we know we're moving a bit ahead of ourselves, but keep the faith), you need to find a small group of potato chip eaters that is representative of all potato chip fans and ask them to tell you which one of the three names they like the most. Then, if you did things right, you can easily infer the findings to the huge group of potato chip eaters.

Or, let's say you're interested in the best treatment for a particular type of disease. Perhaps you'll try a new drug as one alternative, a placebo (or a substance that is known not to have any effect) as another alternative, and even nothing as the third alternative to see what happens. Well, you find out that a larger number of patients get better when no action is taken and nature just takes its course! The drug does not have any effect. Then, with that information, you infer to the larger group of patients that suffers from the disease, given the results of your experiment.

In Other Words . . .

Statistics is a tool that helps us understand the world around us. It does so by organizing information we've collected and then letting us make certain statements about how characteristics of those data are applicable to new settings. Descriptive and inferential statistics work hand in hand, and which one you use and when depends on the question you want answered.

WHAT AM I DOING IN A STATISTICS CLASS?

There are probably many reasons why you find yourself using this book. You might be enrolled in an introductory statistics class. Or you might be reviewing for your comprehensive exams. Or you might even be reading this on summer vacation (horrors!) in preparation and review for a more advanced class.

In any case, you're a statistics student whether you have to take a final exam at the end of a formal course, you're just in it of your own accord, or you're taking the course online 500 miles from the instructor.

But there are plenty of good reasons to be studying this material—some fun, some serious, and some both. Here's the list of some of the things that my students hear at the beginning of our introductory statistics course.

1. Statistics 101 or Statistics 1 or whatever it's called at your school looks great listed on your transcript. Kidding aside, this may be a required course for you to complete your major. But even if it is not, having these skills is definitely a big plus when it comes time to apply for a job or for further schooling. And with more advanced courses, your résumé will be even more impressive. In tough job markets, an edge like this is very important.

2. If this is not a required course, taking basic statistics sets you apart from those students who do not. It shows that you are willing to undertake a course that is above average in difficulty and commitment.

3. Basic statistics is an intellectual challenge of a kind that you might not be used to. There's a good deal of thinking that's required,

a bit of math, and some integration of ideas and application. The bottom line is that all this activity adds up to what can be an invigorating intellectual experience because you learn about a whole new area or discipline.

4. There's no question that having some background in statistics makes you a better student in the social or behavioral sciences because you will have a better understanding not only of what you read in journals but also what your professors and colleagues may be discussing and doing in and out of class. You will be amazed the first time you say to yourself, "Wow, I actually understand what they're talking about." And it will happen over and over again because you will have the basic tools necessary to understand exactly how scientists reach the conclusions they do.

5. If you plan to pursue a graduate degree in education, anthropology, economics, nursing, sociology, or any one of many other social, behavior, and biological pursuits, this course will give you the foundation you need to go further.

6. Finally, you can brag that you completed a course that everyone thinks is the equivalent of building and running a nuclear reactor.

TEN WAYS TO USE THIS BOOK (AND LEARN STATISTICS AT THE SAME TIME!)

Yep. Just what the world needs—another statistics book. But this one is different. It's directed at the student, is not condescending, is informative, and is as basic as possible in its presentation. It makes no presumptions about what you should know before you start and proceeds in slow, small steps, letting you pace yourself.

However, there has always been a general aura surrounding the study of statistics that it's a difficult subject to master. And we don't say otherwise, because parts of it *are* challenging. On the other hand, millions and millions of students have mastered this topic, and you can too. Here are a few hints to close this introductory chapter before we move on to our first topic.

1. **You're not dumb.** That's true. If you were, you would not have gotten this far in school. So treat statistics like any other new course. Attend the lectures, study the material, and do the exercises in the book and from class, and you'll do fine. Rocket scientists know statistics, but you don't have to be a rocket scientist to succeed in statistics. You do have to show up to class and do the homework.

2. **How do you know statistics is hard?** Is statistics difficult? Yes and no. If you listen to friends who have taken the course and didn't work hard and didn't do well, they'll surely volunteer to tell you how hard it was and how much of a disaster it made of their entire semester, if not their life. And let's not forget—we always tend to hear from complainers. So we'd suggest that you start this course with the attitude that you'll wait and see how it is and judge the experience for yourself. Better yet, talk to several people who have had the class and get a good general idea of what they think. Just don't base it on one spoilsport's experience.

3. **Don't skip lessons—work through the chapters in sequence.** *Statistics for People Who (Think They) Hate Statistics* is written so that each chapter provides a foundation for the next one in the book. When you are all done with the course, you will (we hope) refer back to this book and use it as a reference. So, if you need a particular value from a table, you might consult Appendix B. Or if you need to remember how to compute the standard deviation, you might turn to Chapter 3. But for now, read each chapter in the sequence that it appears. It's OK to skip around and see what's offered down the road. Just don't study later chapters before you master earlier ones.

4. **Form a study group.** This is one of the most basic ways to ensure some success in this course. Early in the semester, arrange to study with friends. If you don't have any friends who are in the same class as you, then make some new ones or offer to study with someone who looks to be as happy about being there as you are. Studying with others allows you to help them if you know the material better, or to benefit from others who know that material better than you. Set a specific time each week to get together for an hour and go over the exercises at the end of the chapter or ask questions of one another. Take as much time as you need. Studying with others is an invaluable way to help you understand and master the material in this course.

5. **Ask your teacher questions, and then ask a friend.** If you do not understand what you are being taught in class, ask your professor to clarify it. Have no doubt—if you don't understand the material, then you can be sure that others do not as well. More often than not, instructors welcome questions. And

especially because you've read the material before class, your questions should be well informed and help everyone in class to better understand the material.

6. **Do the exercises at the end of a chapter.** The exercises are based on the material and the examples in the chapter they follow. They are there to help you apply the concepts that were taught in the chapter and build your confidence at the same time. How do the exercises do that? An explanation for how each exercise is solved accompanies the problem. If you can answer these end-of-chapter exercises, then you are well on your way to mastering the content of the chapter.

7. **Practice, practice, practice.** Yes, it's a very old joke:

   Q. How do you get to Carnegie Hall?

   A. Practice, practice, practice.

   Well, it's no different with basic statistics. You have to use what you learn and use it frequently to master the different ideas and techniques. This means doing the exercises in the back of Chapters 1–17 as well as taking advantage of any other opportunities you have to understand what you have learned.

8. **Look for applications to make it more real.** In your other classes, you probably have occasion to read journal articles, talk about the results of research, and generally discuss the importance of the scientific method in your own area of study. These are all opportunities to look and see how your study of statistics can help you better understand the topics under class discussion as well as the area of beginning statistics. The more you apply these new ideas, the better and more full your understanding will be.

9. **Browse.** Read over the assigned chapter first, then go back and read it with more intention. Take a nice leisurely tour of *Statistics for People Who (Think They) Hate Statistics* to see what's contained in the various chapters. Don't rush yourself. It's always good to know what topics lie ahead as well as to familiarize yourself with the content that will be covered in your current statistics class.

10. **Have fun.** This indeed might seem like a strange thing to say, but it all boils down to you mastering this topic rather than letting the course and its demands master you. Set up a study schedule and follow it, ask questions in class, and consider this intellectual exercise to be one of growth. Mastering new material is always exciting and satisfying—it's part of the human spirit. You can experience the same satisfaction here—just keep your eye on the ball and make the necessary commitment to stay current with the assignments and work hard.

ABOUT THOSE ICONS

An icon is a symbol. Throughout *Statistics for People . . .*, you'll see a variety of different icons. Here's what each one is and what each represents:

This icon represents information that goes beyond the regular text. We might find it necessary to elaborate on a particular point, and we can do that more easily outside of the flow of the usual material.

Here, we select some more technical ideas and tips to discuss and to inform you about what's beyond the scope of this course. You might find these interesting and useful.

Throughout *Statistics for People . . .*, you'll find a small steps icon like the one you see here. This indicates that there is a set of steps coming up that will direct you through a particular process. These steps have been tested and approved by whatever federal agency approves these things.

That finger with the bow is a cute icon, but its primary purpose is to help reinforce important points about the topic that you just read about. Try to emphasize these points in your studying because they are usually central to the topic.

Most of the chapters in *Statistics for People . . .*, provide detailed information about one or more particular statistical procedure and the computation that accompanies them. The computer icon is used to identify the "Using the Computer to . . ." section of the chapter.

Many of these chapters also contain instructions for using Version 19.0 of SPSS to complete the same procedures so that you can have both hands-on experience and experience using one of the most powerful statistical analysis packages available today.

Appendix A contains an introduction to SPSS. Working through this appendix is all you really need to do to be ready to use SPSS. If you have an earlier version of SPSS (or the Mac version), you will still find this material to be very helpful. In fact, the latest Windows and Mac versions of SPSS are almost identical in appearance and functionality.

In working the exercises in this book, you will use the data sets in Appendix C, starting on page 366. Also, when you get to each section titled "Using the Computer to . . .," you'll find reference to a data set (such as "Chapter 2 Data Set 1"). Each of these sets is shown in Appendix C, and you will use these data to successfully complete the "Using the Computer to . . ." sections if you want to follow along. You can enter the data manually or download it from either the website hosted by SAGE at http://www.sagepub .com/salkind4e or the author's website at http://www.online filefolder.com with

User name: *ancillaries*

Password: *files*

The data files are available in either SPSS or Excel format.

KEY TO DIFFICULTY INDEX

1.	very hard	☺
2.	hard	☺☺
3.	not too hard, but not easy either	☺☺☺
4.	easy	☺☺☺☺
5.	very easy	☺☺☺☺☺

GLOSSARY

Bolded terms in the text are included in the glossary at the back of the book.

SUMMARY

That couldn't have been that bad, right? We want to encourage you to continue reading and not worry about what's difficult or time consuming or too complex for you to understand and apply. Just take one chapter at a time, as you did this one.

TIME TO PRACTICE

Because there's no substitute for the real thing, Chapters 1–17 end with a set of exercises that will help you review the material that was covered in the chapter.

For example, here is the first set of exercises.

1. Interview someone who uses statistics in his or her everyday work. It might be your adviser, an instructor, a researcher who lives on your block, a market analyst for a company, or even a city planner. Ask them what their first statistics

course was like. Find out what they liked and what they didn't. See if they have any suggestions to help you succeed. And most important, ask the person about the ways he or she uses these new-to-you tools at work.

2. We hope that you are part of a study group, or if that is not possible, that you have a telephone or online study buddy (or even more than one). Talk to your group or a fellow student in your class about similar likes, dislikes, fears, and so on, about the statistics course. What do you have in common? Not in common? Discuss with your fellow student strategies to overcome your fears.

3. Search through your local newspaper and find the results of a survey or interview about any topic. Summarize what the results are and do the best job you can describing how the researchers who were involved, or the authors of the survey, came to the conclusions they did. It may or may not be apparent. Once you have some idea of what they did, try to speculate as to what other ways the same information might be collected, organized, and summarized.

4. Go to the library and copy a journal article in your own discipline. Then go through the article with one of those fancy highlighters and highlight the section (usually the "Results" section) where statistical procedures were used to organize and analyze the data. You don't know much about the specifics of this yet, but how many of these different procedures (such as *t* test, mean, and calculation of the standard deviation) can you identify? Can you take the next step and tell your instructor how the results relate to the research question or the primary topic of the research study?

5. Find five websites on the Internet that contain data on any topic and write a brief description of what type of information is offered and how it is organized. For example, if you go to the mother of all data sites, the United States Census (at http://www.census.gov/), you'll find a link to Data Tools, which takes you to a page just loaded with links to real live data. Try to find data and information that fits in your own discipline.

6. Finally, as your last in this first set of exercises, come up with five of the most interesting questions you can about your own area of study or interest. Do your best to come up with questions for which you would want real, existing information or data. Be a scientist!

PART II

Σigma Freud and Descriptive Statistics

Snapshots

And you thought *your* statistics professor was tough.

One of the things that Sigmund Freud, the founder of psychoanalysis, did quite well was to observe and describe the nature of his patients' conditions. He was an astute observer and used his skills to develop what was the first systematic and comprehensive theory of personality. Regardless of what you may think about the validity of his ideas, he was a good scientist.

Back in the early 20th century, courses in statistics (like the one you are taking) were not offered as part of undergraduate or graduate curricula. The field was relatively new, and the nature of scientific explorations did not demand the precision that this set of tools brings to the scientific arena.

But things have changed. Now, in almost any endeavor, numbers count. This section of *Statistics for People Who (Think They) Hate Statistics* is devoted to understanding how we can use statistics to describe an outcome and better understand it, once the information about the outcome is organized.

Chapter 2 discusses measures of central tendency and how computing one of several different types of averages gives you the one best data point that represents a set of scores. Chapter 3 completes the coverage of tools we need to fully describe a set of data points in its discussion of variability, including the standard deviation and variance. When you get to Chapter 4, you will be ready to learn how distributions, or sets of scores, differ from one another and what this difference means. Chapter 5 deals with the nature of relationships between variables, namely, correlations. And finally, Chapter 6 introduces you to the importance of reliability when we are describing some of the qualities of effective measurement tools.

When you finish Part II, you'll be in excellent shape to start understanding the role that probability and inference play in the social and behavioral sciences.

2 Means to an End

Computing and Understanding Averages

Difficulty Scale ☺☺☺☺ (moderately easy)

WHAT YOU'LL LEARN ABOUT IN THIS CHAPTER

◆ Understanding measures of central tendency

◆ Computing the mean for a set of scores

◆ Computing the mode and the median for a set of scores

◆ Selecting a measure of central tendency

You've been very patient, and now it's finally time to get started working with some real, live data. That's exactly what you'll do in this chapter. Once data are collected, a usual first step is to organize the information using simple indexes to describe the data. The easiest way to do this is through computing an average, of which there are several different types.

An **average** is the one value that best represents an entire group of scores. It doesn't matter whether the group of scores is the number correct on a spelling test for 30 fifth graders or the batting percentage of each of the New York Yankees or the number of people who registered as Democrats or Republicans in the most recent election. In all of these examples, groups of data can be summarized using an average. Averages, also called **measures of central tendency**, come in three flavors: the mean, the median, and the mode. Each provides you with a different type of information about a distribution of scores and is simple to compute and interpret.

COMPUTING THE MEAN

The mean is the most common type of average that is computed. It is simply the sum of all the values in a group, divided by the number of values in that group. So, if you had the spelling scores for 30 fifth graders, you would simply add up all the scores and get a total, and then divide by the number of students, which is 30.

The formula for computing the mean is shown in Formula 2.1.

$$\overline{X} = \frac{\Sigma X}{n} \qquad (2.1)$$

where

- The letter X with a line above it (also sometimes called "X bar") is the mean value of the group of scores or the mean.
- The Σ, or the Greek letter sigma, is the summation sign, which tells you to add together whatever follows it.
- The X is each individual score in the group of scores.
- Finally, the n is the size of the sample from which you are computing the mean.

To compute the mean, follow these steps:

1. List the entire set of values in one or more columns. These are all the Xs.
2. Compute the sum or total of all the values.
3. Divide the total or sum by the number of values.

For example, if you needed to compute the average number of shoppers at three different locations, you would compute a mean for that value.

Location	Number of Annual Customers
Lanham Park store	2,150
Williamsburg store	1,534
Downtown store	3,564

The mean or average number of shoppers in each store is 2,416. Formula 2.2 shows how it was computed using the formula you saw in Formula 2.1:

$$\overline{X} = \frac{\Sigma X}{n} = \frac{2,150 + 1,534 + 3,564}{3} = \frac{7,248}{3} = 2,416 \qquad (2.2)$$

Or, if you needed to compute the average number of students in, Grades kindergarten through 6, you would follow the same procedure.

Grade	Number of Students
Kindergarten	18
1	21
2	24
3	23
4	22
5	24
6	25

The mean or average number of students in each class is 22.43. Formula 2.3 shows how it was computed using the formula you saw in Formula 2.1:

$$\overline{X} = \frac{\Sigma X}{n} = \frac{18 + 21 + 24 + 23 + 22 + 24 + 25}{7} = 22.43 \qquad (2.3)$$

See, we told you it was easy. No big deal.

THINGS TO REMEMBER

The mean is sometimes represented by the letter M and is also called the typical, average, or most central score. If you are reading another statistics book or a research report, and you see something like $M = 45.87$, it probably means that the mean is equal to 45.87.

- In the formula, a small n represents the sample size for which the mean is being computed. A large N (like this) would represent the population size. In some books and in some journal articles, no distinction is made between the two.

- The sample mean is the measure of central tendency that most accurately reflects the population mean.
- The mean is like the fulcrum on a seesaw. It's the centermost point where all the values on one side of the mean are equal in weight to all the values on the other side of the mean.
- Finally, for better or worse, the mean is very sensitive to extreme scores. An extreme score can pull the mean in one or the other direction and make it less representative of the set of scores and less useful as a measure of central tendency. This, of course, all depends on the values for which the mean is being computed. More about this later.

The mean is also referred to as the **arithmetic mean**, and there are other types of means that you may read about, such as the harmonic mean. Those are used in special circumstances but need not concern you here. And if you want to be technical about it, the arithmetic mean (which is the one that we have discussed up to now) is also defined as the point at which the sum of the deviations is equal to zero (whew!). So, if you have scores like 3, 4, and 5 (where the mean is 4), the sum of the deviations about the mean (–1, 0, and +1) is 0.

Remember that the word *average* means only the one measure that best represents a set of scores, and that there are many different types of averages. Which type of average you use depends on the question that you are asking and the type of data that you are trying to summarize.

Computing a Weighted Mean

You've just seen an example of how to compute a simple mean. But there may be situations where you have the occurrence of more than one value and you want to compute a weighted mean. A weighted mean can be easily computed by multiplying the value by the frequency of its occurrence, adding the total of all the products and then dividing by the total number of occurrences. It beats adding up every individual data point.

To compute a weighted mean, follow these steps:

1. List all the values in the sample for which the mean is being computed, such as those shown in the column labeled Value (the value of X) in the following table.

2. List the frequency with which each value occurs.

3. Multiply the value by the frequency, as shown in the third column.

4. Sum all the values in the Value × Frequency column.

5. Divide by the total frequency.

For example, here's a table that organizes the values and frequencies in a flying proficiency test for 100 airline pilots.

Value	Frequency	Value × Frequency
97	4	388
94	11	1,034
92	12	1,104
91	21	1,911
90	30	2,700
89	12	1,068
78	9	702
60 (don't fly with this guy)	1	60
Total	**100**	**8,967**

The weighted mean is 8,967/100, or 89.67. Computing the mean this way is much easier than entering 100 different scores into your calculator or computer program.

In basic statistics, an important distinction needs to be made between those values associated with samples (a part of a population) and those associated with populations. To do this, statisticians use the following conventions. For a sample statistic (such as the mean of a sample), Roman letters are used. For a population parameter (such as the mean of a population), Greek letters are used. So, the mean for the spelling score for a sample of 100 fifth graders is represented as $\overline{X}_5$, whereas the mean for the spelling score for the entire population of fifth graders is represented as μ_5, using the Greek letter mu, or μ.

COMPUTING THE MEDIAN

The median is also an average but of a very different kind. The **median** is defined as the midpoint in a set of scores. It's the point at which one half, or 50%, of the scores fall above and one half, or 50%, fall below. It's got some special qualities that we will talk about later in this section, but for now, let's concentrate on how it is computed. There's no standard formula for computing the median.

To compute the median, follow these steps:

1. List the values in order, either from highest to lowest or lowest to highest.

2. Find the middle-most score. That's the median.

For example, here are the incomes from five different households:

$135,456

$25,500

$32,456

$54,365

$37,668

Here is the list ordered from highest to lowest:

$135,456

$54,365

$37,668

$32,456

$25,500

There are five values. The middle-most value is $37,668, and that's the median.

Now, what if the number of values is even? Let's add a value ($34,500) to the list so there are six income levels. Here they are.

$135,456

$54,365

$37,668

$34,500

$32,456

$25,500

When there is an even number of values, the median is simply the mean between the two middle values. In this case, the middle two cases are $34,500 and $37,668. The mean of those two values is $36,084. That's the median for that set of six values.

What if the two middle-most values are the same, such as in the following set of data?

$45,678

$25,567

$25,567

$13,234

Then the median is same as both of those middle-most values. In this example, it's $25,567.

If we had a series of values that was the number of days spent in rehabilitation for a sports-related injury for seven different patients, the numbers may look like this:

43

34

32

12

51

6

27

As we did before, we can order the values (51, 43, 34, 32, 27, 12, 6) and then select the middle value as the median, which in this case is 32. So, the median number of days spent in rehab is 32.

Tech Talk

If you know about medians, you should know about **percentile points**. Percentile points are used to define the percentage of cases equal to and below a certain point in a distribution or set of scores. For example, if a score is "at the 75th percentile," it means that the score is at or above 75% of the other scores in the distribution. The median is also known as the 50th percentile, because it's the point below which 50% of the cases in the distribution fall. Other percentiles are useful as well, such as the 25th percentile, often called Q_1, and the 75th percentile, referred to as Q_3. So what's Q_2? The median, of course.

Here comes the answer to the question you've probably had in the back of your mind since we started talking about the median. Why use the median instead of the mean? For one very good reason. The median is insensitive to extreme scores, whereas the mean is not.

When you have a set of scores in which one or more scores are extreme, the median better represents the centermost value of that set of scores than any other measure of central tendency. Yes, even better than the mean.

What do we mean by extreme? It's probably easiest to think of an extreme score as one that is very different from the group to which it belongs. For example, consider the list of five incomes that we worked with earlier (shown again here):

$135,456

$54,365

$37,668

$32,456

$25,500

The value $135,456 is more different from the other five than any other value in the set. We would consider that an extreme score.

The best way to illustrate how useful the median is as a measure of central tendency is to compute both the mean and the median for a set of data that contains one or more extreme scores and then compare them to see which one best represents the group. Here goes.

The mean of the set of five scores you see above is the sum of the set of five divided by five, which turns out to be $57,089. On the other hand, the median for this set of five scores is $37,668. Which is more representative of the group? The value $37,668, because it clearly lies more in the middle of the group, and we like to think about "the average" (in this case, we are using the median as a measure of average) as being representative or assuming a central position. In fact, the mean value of $57,089 falls above the fourth highest value ($54,365) and is not very central or representative of the distribution.

It's for this reason that certain social and economic indicators (mostly involving income) are reported using a median as a measure of central tendency, such as "The median income of the average American family is . . . ," rather than using the mean to summarize the values. There are just too many extreme scores that would **skew**, or significantly distort, what is actually a central point in the set or distribution of scores.

> You learned earlier that sometimes the mean is represented by the capital letter *M* instead of $\overline{X}$. Well, other symbols are used for the median as well. We like the letter *M*, but some people confuse it with the mean, so they use Med for median, or Mdn. Don't let that throw you—just remember what the median is and what it represents, and you'll have no trouble adapting to different symbols.

THINGS TO REMEMBER

Here are some interesting and important things to remember about the median.

- The mean is the middle point of a set of values, and the median is the middle point of a set of cases.
- Because the median cares about how many cases, and not the values of those cases, extreme scores (sometimes called **outliers**) don't count.

COMPUTING THE MODE

The third and last measure of central tendency that we'll cover, the mode, is the most general and least precise measure of central tendency, but it plays a very important part in understanding the characteristics of a special set of scores. The **mode** is the value that occurs most frequently. There is no formula for computing the mode.

To compute the mode, follow these steps:

1. List all the values in a distribution, but list each only once.
2. Tally the number of times that each value occurs.
3. The value that occurs most often is the mode.

For example, an examination of the political party affiliation of 300 people might result in the following distribution of scores.

Party Affiliation	Number or Frequency
Democrats	90
Republicans	70
Independents	140

The mode is the value that occurs most frequently, which in the above example is Independents. That's the mode for this distribution.

If we were looking at the modal response on a 100-item multiple-choice test, we might find that the A alternative was chosen more frequently than any other. The data might look like this.

Item Alternative Selected	A	B	C	D
Number of Times	57	20	12	11

On this 100-item multiple-choice test where each item has four choices (A, B, C, and D), A was the answer selected 57 times. It's the modal response.

Want to know what the easiest and most commonly made mistake is when computing the mode? It's selecting the number of times a category occurs, rather than the label of the category itself. Instead of the mode being Independents, it's easy for someone to conclude the mode is 140. Why? Because they are looking at the number of times the value occurred, and not the value that occurred most often! This is a simple mistake to make, so be on your toes when you are asked about these things.

Apple Pie à la Bimodal

If every value in a distribution contains the same number of occurrences, then there really isn't a mode. But if more than one value appears with equal frequency, the distribution is multimodal. The set of scores can be bimodal (with two modes), as the following set of data using hair color illustrates.

Hair Color	Number or Frequency
Red	7
Blond	12
Black	45
Brown	45

In the above example, the distribution is bimodal because the frequency of the values of black and brown hair occurs equally. You can even have a bimodal distribution when the modes are relatively close together, but not exactly the same, such as 45 people with black hair and 44 with brown hair. The question becomes, How much does one class of occurrences stand apart from another? Can you have a trimodal distribution? Sure—where three values have the same frequency. It's unlikely, especially when you are dealing with a large set of **data points**, or observations, but certainly possible. The real answer to the above stand-apart question is that categories have to be mutually exclusive—you simply cannot have both black and red hair (although if you look around the classroom, you may think differently). Of course, you can have those two colors, but each person's hair color is forced into only one category.

WHEN TO USE WHAT

OK, we've defined three different measures of central tendency and given you fairly clear examples of each. But the most important question remains unanswered. That is, "When do you use which measure?"

In general, which measure of central tendency you use depends on the type of data that you are describing. Unquestionably, a measure of central tendency for qualitative, categorical, or nominal data (such as racial group, eye color, income bracket, voting preference, and neighborhood location) can be described using only the mode.

For example, you can't be looking at the most central measure that describes which political affiliation is most predominant in a group and use the mean—what in the world could you conclude, that everyone is half-Republican? Rather, that out of 300 people, almost half (140) are Independent seems to be the best way of describing the value of this variable. In general, the median and mean are best used with quantitative data, such as height, income level in dollars (not categories), age, test score, reaction, and number of hours completed for a degree.

It's also fair to say that the mean is a more precise measure than the median, and the median is a more precise measure than the mode. This means that all other things being equal, use the mean, and indeed, the mean is the most often used measure of central tendency. However, we do have occasions when the mean would not be appropriate as a measure of central tendency—for example, when we have categorical or nominal data, such as hospitalized people. Then we use the mode. So, here is a set of three guidelines that may be of some help. And remember, there can always be exceptions.

1. Use the mode when the data are categorical in nature and values can fit into only one class, such as hair color, political affiliation, neighborhood location, and religion. When this is the case, these categories are called mutually exclusive.

2. Use the median when you have extreme scores and you don't want to distort the average (computed as the mean), such as when the variable of interest is income expressed in dollars.

3. Finally, use the mean when you have data that do not include extreme scores and are not categorical, such as the numerical score on a test or the number of seconds it takes to swim 50 yards.

USING THE COMPUTER AND COMPUTING DESCRIPTIVE STATISTICS

If you haven't already, now would be a good time to turn to Appendix A so you can become familiar with the basics of using SPSS. Then come back here.

Let's use SPSS to compute some descriptive statistics. The data set we are using is named Chapter 2 Data Set 1, which is a set of 20 scores on a test of prejudice. All of the data sets are available in Appendix C and from the SAGE website (www.sagepub.com/salkind4e) or from www.onlinefilefolder.com (see page 15 for instructions about this location on the Internet) or from the author at njs@ku.edu. There is one variable in this data set:

Variable	Definition
Prejudice	The value on a test of prejudice as measured on a scale from 1 to 100

Here are the steps to compute the measures of central tendency that we discussed in this chapter. Follow along and do it yourself. With this and all exercises, including data that you enter or download, we'll assume that the data set is already open in SPSS.

1. Click Analyze → Descriptive Statistics → Frequencies.

2. Double-click on the variable named Prejudice to move it to the Variable(s) box.

3. Click Statistics and you will see the Frequencies: Statistics dialog box shown in Figure 2.1.

Figure 2.1 The Frequencies: Dialog Box From SPSS

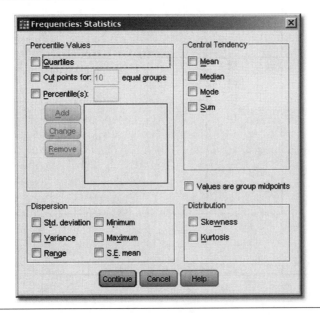

4. Under Central Tendency, click the Mean, Median, and Mode boxes.

5. Click Continue.

6. Click OK.

The SPSS Output

Figure 2.2 shows you selected output from the SPSS procedure for the variable named Prejudice.

In the Statistics part of the output, you can see how the mean, median, and mode are all computed along with the sample size and the fact that there were no missing data. SPSS does not use symbols such as X in its output. Also listed in the output are the frequency of each value and the percentage of times it occurs, all useful descriptive information.

| Figure 2.2 | Descriptive Statistics From SPSS |

Statistics

Predjudice

N	Valid	20
	Missing	0
Mean		84.70
Median		87.00
Mode		87

Predjudice

		Frequency	Percent	Valid Percent	Cumulative Percent
Valid	55	1	5.0	5.0	5.0
	64	1	5.0	5.0	10.0
	67	1	5.0	5.0	15.0
	76	1	5.0	5.0	20.0
	77	1	5.0	5.0	25.0
	81	2	10.0	10.0	35.0
	82	1	5.0	5.0	40.0
	87	4	20.0	20.0	60.0
	89	1	5.0	5.0	65.0
	93	1	5.0	5.0	70.0
	94	2	10.0	10.0	80.0
	96	1	5.0	5.0	85.0
	99	3	15.0	15.0	100.0
	Total	20	100.0	100.0	

It's a bit strange, but if you select Analyze → Descriptive Statistics → Descriptives in SPSS and then click Options, there's no option to select the median or the mode, which you might expect because they are basic descriptive statistics. The lesson here? Statistical analysis programs are usually quite different from one another, use different names for the same things, and make different assumptions about what's where. If you can't find what you want, it's probably there. Just keep hunting. Also, be sure to use the Help feature to help navigate through all this new information until you find what you need.

SUMMARY

No matter how fancy schmancy your statistical techniques are, you will still almost always start by simply describing what's there—hence the importance of understanding the simple notion of central tendency. From here, we go to another important descriptive construct: variability, or how different scores are from one another.

TIME TO PRACTICE

1. By hand, compute the mean, median, and mode for the following set of 40 reading scores.

31	32	43	42
24	34	25	44
23	43	24	36
25	41	23	28
14	21	24	17
25	23	44	21
13	26	23	32
12	26	14	42
14	31	52	12
23	42	32	34

2. Compute the mean, median, and mode for the following three sets of scores saved as Chapter 2 Data Set 2. Do it by hand or use a computer program such as SPSS. Show your work, and if you use SPSS, print out a copy of the output.

Score 1	Score 2	Score 3
3	34	154
7	54	167
5	17	132
4	26	145
5	34	154
6	25	145
7	14	113
8	24	156
6	25	154
5	23	123

3. Compute the means for the following set of scores saved as Chapter 2 Data Set 3 using SPSS. Print out a copy of the output.

Hospital Size (number of beds)	Infection Rate (per 1,000 admissions)
234	1.7
214	2.4
165	3.1
436	5.6
432	4.9
342	5.3
276	5.6
187	1.2
512	3.3
553	4.1

4. You are the manager of a fast food store. Part of your job is to report to the boss at the end of each day which special is selling best. Use your vast knowledge of descriptive statistics and write one paragraph to let the boss know what happened today. Here are the data. Don't use SPSS to compute important values; rather, do it by hand. Be sure to include a copy of your work.

Special	Number Sold	Cost
Huge Burger	20	$2.95
Baby Burger	18	$1.49
Chicken Littles	25	$3.50
Porker Burger	19	$2.95
Yummy Burger	17	$1.99
Coney Dog	20	$1.99
Total specials sold	119	

5. Imagine yourself as the CEO of a huge corporation and you are planning an expansion. You'd like your new store to have some of the same typical numbers as the other three that are under your empire. By hand, provide some idea of what you want the stores to look like. And, remember that you have to select whether to use the mean, median, or mode as an average. Good luck, young Jedi.

Average	Store 1	Store 2	Store 3	Store 4	New Store
Sales (in thousands)	323.6	234.6	308.3	?	
Number of items purchased	3,454	5,645	4,565	?	
Number of visitors	4,534	6,765	6,654	?	

6. Under what conditions would you use the median rather than the mean as a measure of central tendency? Why? Provide an example of two situations where the median might be more useful than the mean as a measure of central tendency.

7. Suppose you are working with a data set that has some very "different" (much larger or much smaller than the rest of the data) scores. What measure of central tendency would you use and why?

8. For this exercise, use the following set of 16 scores (ranked) that consists of income levels ranging from about $50,000 to about $200,000. What is the best measure of central tendency and why?

$199,999

$98,789

$90,878

$87,678

$87,245

$83,675

$77,876

$77,743

$76,564

$76,465

$75,643

$66,768

$65,654

$58,768

$54,678

$51,354

Vive la Différence

Understanding Variability

Difficulty Scale ☺☺☺☺ (moderately easy, but not a cinch)

✦ Why variability is valuable as a descriptive tool

✦ How to compute the range, standard deviation, and variance

✦ How the standard deviation and variance are alike—and how they are different

WHY UNDERSTANDING VARIABILITY IS IMPORTANT

In Chapter 2, you learned about different types of averages, what they mean, how they are computed, and when to use them. But when it comes to descriptive statistics and describing the characteristics of a distribution, averages are only half the story. The other half is measures of variability.

In the most simple of terms, **variability** reflects how scores differ from one another. For example, the following set of scores shows some variability:

$$7, 6, 3, 3, 1$$

The following set of scores has the same mean (4) and has less variability than the previous set:

$$3, 4, 4, 5, 4$$

The next set has no variability at all—the scores do not differ from one another—but it also has the same mean as the other two sets we just showed you.

$$4, 4, 4, 4, 4$$

Variability (also called spread or dispersion) can be thought of as a measure of how different scores are from one another. It's even more accurate (and maybe even easier) to think of variability as how different scores are from one particular score. And what "score" do you think that might be? Well, instead of comparing each score to every other score in a distribution, the one score that could be used as a comparison is—that's right—the mean. So, variability becomes a measure of how much each score in a group of scores differs from the mean. More about this in a moment.

Remember what you already know about computing averages—that an average (whether it is the mean, the median, or the mode) is a representative score in a set of scores. Now, add your new knowledge about variability—that it reflects how different scores are from one another. Each is an important descriptive statistic. Together, these two (average and variability) can be used to describe the characteristics of a distribution and show how distributions differ from one another.

Three measures of variability are commonly used to reflect the degree of variability, spread, or dispersion in a group of scores. These are the range, the standard deviation, and the variance. Let's take a closer look at each one and how each one is used.

Actually, how data points differ from one another is a central part of understanding and using basic statistics. But when it comes to differences between individuals and groups (a mainstay of most social and behavior sciences), the whole concept of variability becomes really important. Sometimes it's called fluctuation, or lability, or error, or one of many other terms, but the fact is, variety is the spice of life, and what makes people different from one another also makes understanding them and their behavior all the more challenging (and interesting). Without variability either in a set of data or between individuals and groups, things are just boring.

COMPUTING THE RANGE

The range is the most general measure of variability. It gives you an idea of how far apart scores are from one another. The **range** is computed simply by subtracting the lowest score in a distribution from the highest score in the distribution.

In general, the formula for the range is

$$r = h - l \qquad\qquad (3.1)$$

where

r   is the range

h   is the highest score in the data set

l   is the lowest score in the data set

Take the following set of scores, for example (shown here in descending order):

$$98, 86, 77, 56, 48$$

In this example, 98 − 48 = 50. The range is 50. In a set of 500 numbers, where the largest is 98 and the smallest is 37, then the range would be 61.

There really are two kinds of ranges. One is the *exclusive range,* which is the highest score minus the lowest score (or $h - l$) and the one we just defined. The second kind of range is the *inclusive range,* which is the highest score minus the lowest score plus 1 (or $h - l + 1$). You most commonly see the exclusive range in research articles, but the inclusive range is also used on occasion if the researcher prefers it.

The range is used almost exclusively to get a very *general* estimate of how wide or different scores are from one another—that is, the range shows how much spread there is from the lowest to the highest point in a distribution.

So, although the range is fine as a general indicator of variability, it should not be used to reach any conclusions regarding how individual scores differ from one another. And, you will usually never see it reported as the only measure of variability, but as one of several (which brings us to . . .)

COMPUTING THE STANDARD DEVIATION

Now we get to the most frequently used measure of variability, the standard deviation. Just think about what the term implies; it's a deviation from something (guess what?) that is standard. Actually, the **standard deviation** (abbreviated as *s* or *SD*) represents the average amount of variability in a set of scores. In practical terms, it's the average distance from the mean. The larger the standard deviation, the larger the average distance each data point is from the mean of the distribution.

So, what's the logic behind computing the standard deviation? Your initial thoughts may be to compute the mean of a set of scores and then subtract each individual score from the mean. Then, compute the average of that distance.

That's a good idea—you'll end up with the average distance of each score from the mean. But it won't work (see if you know why even though we'll show you why in a moment).

First, here's the formula for computing the standard deviation:

$$s = \sqrt{\frac{\Sigma(X - \overline{X})^2}{n - 1}} \qquad (3.2)$$

where

    s   is the standard deviation

    Σ   is sigma, which tells you to find the sum of what follows

    X   is each individual score

    $\overline{X}$ is the mean of all the scores

    n   is the sample size.

This formula finds the difference between each individual score and the mean $(X - \overline{X})$, squares each difference, and sums them all together. Then, it divides the sum by the size of the sample (minus 1) and takes the square root of the result. As you can see, and as we mentioned earlier, the standard deviation is an average deviation from the mean.

Here are the data we'll use in the following step-by-step explanation of how to compute the standard deviation.

5, 8, 5, 4, 6, 7, 8, 8, 3, 6

1. List each score. It doesn't matter whether the scores are in any particular order.

2. Compute the mean of the group.

3. Subtract the mean from each score.

Here's what we've done so far, where $X - \overline{X}$ represents the difference between the actual score and the mean of all the scores, which is 6.

X	$\overline{X}$	$X - \overline{X}$
8	6	8 − 6 = +2
8	6	8 − 6 = +2
8	6	8 − 6 = +2
7	6	7 − 6 = +1
6	6	6 − 6 = 0
6	6	6 − 6 = 0
5	6	5 − 6 = −1
5	6	5 − 6 = −1
4	6	4 − 6 = −2
3	6	3 − 6 = −3

4. Square each individual difference. The result is the column marked $(X - \overline{X})^2$.

X	$(X - \overline{X})$	$(X - \overline{X})^2$
8	+2	4
8	+2	4
8	+2	4
7	+1	1
6	0	0
6	0	0
5	−1	1
5	−1	1
4	−2	4
3	−3	9
Sum	0	28

5. Sum all the squared deviations about the mean. As you can see, the total is 28.

6. Divide the sum by $n - 1$, or $10 - 1 = 9$, so then $28/9 = 3.11$.

7. Compute the square root of 3.11, which is 1.76 (after rounding). That is the standard deviation for this set of 10 scores.

What we now know from these results is that each score in this distribution differs from the mean by an average of 1.76 points.

Let's take a short step back and examine some of the operations in the standard deviation formula. They're important to review and will increase your understanding of what the standard deviation is.

First, why didn't we just add up the deviations from the mean? Because the sum of the deviations from the mean is always equal to 0. Try it by summing the deviations $(2 + 2 + 2 + 1 + 0 + 0 - 1 - 1 - 2 - 3)$. In fact, that's the best way to check if you computed the mean correctly.

Tech Talk

There's another type of deviation that you may read about, and you should know what it means. The **mean deviation** (also called the mean absolute deviation) is the sum of the absolute value of the deviations from the mean divided by the number of scores. You already know that the sum of the deviations from the mean must equal 0 (otherwise the mean is probably computed incorrectly). Instead, let's take the absolute value of each deviation (which is the value regardless of the sign). Sum them together and divide by the number of data points, and you have the mean deviation. So, if you have a set of scores such as 3, 4, 5, 5, 8, and the arithmetic mean is 5, the mean deviation is 2 (the absolute value of $5 - 3$), 1, 0, 0, and 3, for a total of 6. (Note: The absolute value of a number is usually represented as that number with a vertical line on each side of it, such as $|5|$. For example, the absolute value of -6, or $|-6|$, is 6.)

Second, why do we square the deviations? Because we want to get rid of the negative sign so that when we do eventually sum them, they don't add up to 0.

And finally, why do we eventually end up taking the square root of the entire value in Step 7? Because we want to return to the same units with which we originally started. We squared the deviations from the mean in Step 4 (to get rid of negative values) and then took the square root of their total in Step 7. Pretty tidy.

Why n – 1? What's Wrong With Just n?

You might have guessed why we square the deviations about the mean and why we go back and take the square root of their sum. But how about subtracting the value of 1 from the denominator of the formula? Why do we divide by $n-1$ rather than just plain ol' n? Good question.

The answer is that s (the standard deviation) is an estimate of the population standard deviation, and is an **unbiased estimate** at that, but only when we subtract 1 from n. By subtracting 1 from the denominator, we artificially force the standard deviation to be larger than it would be otherwise. Why would we want to do that? Because, as good scientists, we are conservative. Being conservative means that if we have to err, we will do so on the side of overestimating what the standard deviation of the population is. Dividing by a smaller denominator lets us do so. Thus, instead of dividing by 10, we divide by 9. Or instead of dividing by 100, we divide by 99.

Biased estimates are appropriate if your intent is only to describe the characteristics of the sample. But if you intend to use the sample as an estimate of a population parameter, then the unbiased statistic is best to calculate.

Take a look in the following table and see what happens as the size of the sample gets larger (and moves closer to the population in size). The $n-1$ adjustment has far less of an impact on the difference between the biased and the unbiased estimates of the standard deviation (the bold column in the table). All other things being equal, then, the larger the size of the sample, the less of a difference there is between the biased and the unbiased estimates of the standard deviation. Check out the following table, and you'll see what we mean.

Sample Size	Value of Numerator in Standard Deviation Formula	Biased Estimate of the Population Standard Deviation (dividing by n)	Unbiased Estimate of the Population Standard Deviation (dividing by $n-1$)	Difference Between Biased and Unbiased Estimates
10	500	7.07	7.45	.38
100	500	2.24	2.25	.01
1,000	500	0.7071	0.7075	.0004

The moral of the story? When you compute the standard deviation for a sample, which is an estimate of the population, the closer to the size of the population the sample is, the more accurate the estimate will be.

What's the Big Deal?

The computation of the standard deviation is very straightforward. But what does it mean? As a measure of variability, all it tells us is how much each score in a set of scores, on the average, varies from the mean. But it has some very practical applications, as you will find out in Chapter 4. Just to whet your appetite, consider this: The standard deviation can be used to help us compare scores from different distributions, *even when the means and standard deviations are different.* Amazing! This, as you will see, can be very cool.

THINGS TO REMEMBER

- The standard deviation is computed as the average distance from the mean. So, you will need to first compute the mean as a measure of central tendency. Don't fool around with the median or the mode in trying to compute the standard deviation.
- The larger the standard deviation, the more spread out the values are, and the more different they are from one another.
- Just like the mean, the standard deviation is sensitive to extreme scores. When you are computing the standard deviation of a sample and you have extreme scores, note that somewhere in your written report.
- If $s = 0$, there is absolutely no variability in the set of scores, and the scores are essentially identical in value. This will rarely happen.

COMPUTING THE VARIANCE

Here comes another measure of variability and a nice surprise. If you know the standard deviation of a set of scores and you can square a number, you can easily compute the variance of that same set of scores. This third measure of variability, the variance, is simply the standard deviation squared.

In other words, it's the same formula you saw earlier but without the square root bracket, like the one shown in Formula 3.3:

$$s^2 = \frac{\Sigma(X - \overline{X})^2}{n - 1} \qquad (3.3)$$

If you take the standard deviation and never complete the last step (taking the square root), you have the variance. In other words, $s^2 = s \times s$, or the variance equals the standard deviation times itself (or squared). In our earlier example, where the standard deviation was equal to 1.76, the variance is equal to 1.76^2 or 3.11. As another example, let's say that the standard deviation of a set of 150 scores is 2.34. Then, the variance would be 2.34^2 or 5.48.

You are not likely to see the variance mentioned by itself in a journal article or see it used as a descriptive statistic. This is because the variance is a difficult number to interpret and apply to a set of data. After all, it is based on squared deviation scores.

But the variance is important because it is used both as a concept and as a practical measure of variability in many statistical formulas and techniques. You will learn about these later in *Statistics for People Who (Think They) Hate Statistics*.

The Standard Deviation Versus the Variance

How are standard deviation and the variance the same, and how are they different?

Well, they are both measures of variability, dispersion, or spread. The formulas used to compute them are very similar. You see them all over the place in the "Results" sections of journals.

They are also quite different.

First, and most important, the standard deviation (because we take the square root of the average summed squared deviation) is stated in the original units from which it was derived. The variance is stated in units that are squared (the square root is never taken).

What does this mean? Let's say that we need to know the variability of a group of production workers assembling circuit boards. Let's say that they average 8.6 boards per hour, and the standard deviation is 1.59. The value 1.59 means that the difference in the average number of boards assembled per hour is about 1.59 circuit boards from the mean.

Let's look at an interpretation of the variance, which is 1.59^2, or 2.53. This would be interpreted as meaning that the average difference between the workers is about 2.53 circuit boards *squared* from the mean. Which of these two makes more sense?

USING THE COMPUTER TO COMPUTE MEASURES OF VARIABILITY

Let's use SPSS to compute some measures of variability. We are using the file named Chapter 3 Data Set 1.

There is one variable in this data set:

Variable	Definition
ReactionTime	Reaction time on a tapping task

Here are the steps to compute the measures of variability that we discussed in this chapter.

1. Open the file named Chapter 3 Data Set 1.

2. Click Analyze → Descriptive Statistics → Frequencies.

3. Double-click on the ReactionTime variable to move it to the Variable(s) box.

4. Click Statistics, and you will see the Frequencies: Statistics dialog box. This dialog box is used to select the variables and procedures you want to perform.

5. Under Dispersion, click Std. Deviation.

6. Under Dispersion, click Variance.

7. Under Dispersion, click Range.

8. Click Continue.

9. Click OK.

The SPSS Output

Figure 3.1 shows selected output from the SPSS procedure for
ReactionTime. There are 30 valid cases with no missing cases, and
the standard deviation is .70255. The variance equals .494 (or s^2),
and the range is 2.60.

Figure 3.1	Output for the Variable ReactionTime

Statistics

ReactionTime

N	Valid	30
	Missing	0
Std. Deviation		.70255
Variance		.494
Range		2.60

Let's try another one, titled Chapter 3 Data Set 2. There are two
variables in this data set:

Variable	Definition
MathScore	Score on a mathematics test
ReadingScore	Score on a reading test

Follow the same set of instructions as given previously, only in
Step 3, you select both variables. The SPSS output is shown in
Figure 3.2, where you can see selected output from the SPSS procedure for these two variables. There are 30 valid cases with no missing cases, and the standard deviation for math scores is 12.36 with
a variance of 152.7 and a range of 43. For reading scores, the standard deviation is 18.700, the variance is a whopping 349.689 (that's
pretty big), and the range is 76 (which is large as well, reflecting the
similarly large variance).

Figure 3.2   Output for the Variables MathScore and ReadingScore

Statistics

		Math_Score	Reading_Score
N	Valid	30	30
	Missing	0	0
Std. Deviation		12.357	18.700
Variance		152.700	349.689
Range		43	76

SUMMARY

Measures of variability help us even more fully understand what a distribution of data points looks like. Along with a measure of central tendency, we can use these values to distinguish distributions from one another and effectively describe what a collection of test scores, heights, or measures of personality looks like. Now that we can think and talk about distributions, let's explore ways we can look at them.

TIME TO PRACTICE

1. Why is the range the most convenient measure of dispersion, yet the most imprecise measure of variability? When would you use the range?

2. Compute the exclusive and inclusive ranges for the following items.

High Score	Low Score	Inclusive Range	Exclusive Range
7	6		
89	45		
34	17		
15	2		
1	1		

3. Why would you expect more variability on a measure of personality in college freshman graders than you would on a measure of height?

4. Why does the standard deviation get smaller as the individuals in a group score more similarly on a test?

5. For the following set of scores, compute the range, the unbiased and the biased standard deviations, and the variance. Do the exercise by hand.

31, 42, 35, 55, 54, 34, 25, 44, 35

Why is the unbiased estimate greater than the biased estimate?

6. Use SPSS to compute all the descriptive statistics for the following set of three test scores over the course of a semester. Which test had the highest average score? Which test had the smallest amount of variability?

Test 1	Test 2	Test 3
50	50	49
48	49	47
51	51	51
46	46	55
49	48	55
48	53	45
49	49	47
49	52	45
50	48	46
50	55	53

7. For the following set of scores, compute by hand the unbiased estimates of the standard deviation and variance.

4, 5, 6, 2, 5, 7, 5, 6, 8, 5

8. The variance for a set of scores is 25. What is the standard deviation and what is the range?

9. Find the range, standard deviation, and variance of each of the following sets of scores:
   a. 3, 5, 7, 9
   b. .2, .4, .6, .8
   c. 3.5, 6.2, 9.3, 4.1, 5.5, 7.9

10. This practice problem uses the data contained in the file named Chapter 3 Data Set 3. There are two variables in this data set.

Variable	Definition
Height	height in inches
Weight	weight in pounds

Using SPSS, compute all of the measures of variability you can for height and weight.

11. How can you tell whether SPSS produces a biased or an unbiased estimate of the standard deviation?

4

A Picture Really Is Worth a Thousand Words

Difficulty Scale ☺☺☺☺
(moderately easy, but not a cinch)

WHAT YOU'LL LEARN ABOUT IN THIS CHAPTER

✦ Why a picture is really worth a thousand words

✦ How to create a histogram and polygon

✦ Different types of charts and their uses

✦ Using Excel and SPSS to create charts

WHY ILLUSTRATE DATA?

In the previous two chapters, you learned about two important types of descriptive statistics—measures of central tendency and measures of variability. Both of these provide you with the one best score for describing a group of data (central tendency) and a measure of how diverse, or different, scores are from one another (variability).

What we did not do, and what we will do here, is examine how differences in these two measures result in different-looking distributions. Numbers alone (such as $= 10$ and $s = 3$) may be important, but a visual representation is a much more effective way of examining the characteristics of a distribution as well as the characteristics of any set of data.

So, in this chapter, we'll learn how to visually represent a distribution of scores as well as how to use different types of graphs to represent different types of data.

TEN WAYS TO A GREAT FIGURE (EAT LESS AND EXERCISE MORE?)

Whether you create illustrations by hand or use a computer program, the principles of decent design still apply. Here are 10 to copy and put above your desk.

1. **Minimize chart or graph junk.** "Chart junk" (a close cousin to "word junk") is where you use every function, every graph, and every feature a computer program has to make your charts busy, full, and uninformative. Less is definitely more.

2. **Plan out your chart before you start creating the final copy.** Use graph paper even if you will be using a computer program to generate the graph.

3. **Say what you mean and mean what you say—no more and no less.** There's nothing worse than a cluttered (with too much text and fancy features) graph to confuse the reader.

4. **Label everything so nothing is left to the misunderstanding of the audience.**

5. **A graph should communicate only one idea.**

6. **Keep things balanced.** When you construct a graph, center titles and axis labels.

7. **Maintain the scale in a graph.** The scale refers to the relationship between the horizontal and vertical axes. This ratio should be about three to four, so a graph that is three inches wide will be about four inches tall.

8. **Simple is best and less is more.** Keep the chart simple, but not simplistic. Convey the one idea as straightforwardly as possible, with distracting information saved for the accompanying text. Remember, a chart or graph should be able to stand alone, and the reader should be able to understand the message.

9. **Limit the number of words you use.** Too many words, or words that are too large, can detract from the visual message your chart should convey.

10. **A chart alone should convey what you want to say.** If it doesn't, go back to your plan and try it again.

FIRST THINGS FIRST: CREATING A FREQUENCY DISTRIBUTION

The most basic way to illustrate data is through the creation of a frequency distribution. A **frequency distribution** is a method of tallying and representing how often certain scores occur. In the creation of a frequency distribution, scores are usually grouped into class intervals, or ranges of numbers.

Here are 50 scores on a test of reading comprehension and what the frequency distribution for these scores looks like.

Here are the raw data on which it is based:

47	10	31	25	20
2	11	31	25	21
44	14	15	26	21
41	14	16	26	21
7	30	17	27	24
6	30	16	29	24
35	32	15	29	23
38	33	19	28	20
35	34	18	29	21
36	32	16	27	20

And here's the frequency distribution:

Class Interval	Frequency
45–49	1
40–44	2
35–39	4
30–34	8
25–29	10
20–24	10
15–19	8
10–14	4
5–9	2
0–4	1

The Classiest of Intervals

As you can see from the above table, a **class interval** is a range of numbers, and the first step in the creation of a frequency distribution is to define how large each interval will be. As you can see in the frequency distribution that we created, each interval spans five possible scores such as 5–9 (which includes scores 5, 6, 7, 8, and 9) and 40–44 (which includes scores 40, 41, 42, 43, and 44). How did we decide to have an interval that includes only five scores? Why not five intervals, each consisting of 10 scores? Or two intervals, each consisting of 25 scores?

Here are some general rules to follow in the creation of a class interval, regardless of the size of values in the data set you are dealing with.

1. Select a class interval that has a range of 2, 5, 10, 15, or 20 data points. In our example, we chose 5.

2. Select a class interval so that 10 to 20 such intervals cover the entire range of data. A convenient way to do this is to compute the range, then divide by a number that represents the number of intervals you want to use (between 10 and 20). In our example, there are 50 scores and we wanted 10 intervals: 50/10 = 5, which is the size of each class interval. If you had a set of scores ranging from 100 to 400, you can start with the following estimate and work from there: 300/20 = 15, so 15 would be the class interval.

3. Begin listing the class interval with a multiple of that interval. In our frequency distribution shown earlier, the class interval is 5 and we started with the lowest class interval of 0.

4. Finally, the largest interval goes at the top of the frequency distribution.

Once class intervals are created, it's time to complete the frequency part of the frequency distribution. That's simply counting the number of times a score occurs in the raw data and entering that number in each of the class intervals represented by the count.

In the frequency distribution that we created on page 50, the number of scores that occur between 30 and 34 and are in the 30–34 class interval is 8. So, an 8 goes in the column marked Frequency. There's your frequency distribution.

THE PLOT THICKENS: CREATING A HISTOGRAM

Now that we've got a tally of how many scores fall in what class intervals, we'll go to the next step and create what is called a **histogram**, a visual representation of the frequency distribution where the frequencies are represented by bars.

Depending on the book you read and the software you use, visual representations of data are called graphs (such as in SPSS) or charts (such as in the Microsoft spreadsheet Excel). It really makes no difference. All you need to know is that a graph or a chart is the visual representation of data.

To create a histogram, do the following:

1. Using a piece of graph paper, place values at equal distances along the *x*-axis, as shown in Figure 4.1. Now, identify the midpoint of the class intervals, which is the middle point in the class interval. It's pretty easy to just eyeball, but you can also just add the top and bottom values of the class interval and divide by 2. For example, the midpoint of the class interval 0–4 is the average of 0 and 4, or 4/2 = 2.

Figure 4.1 Class Intervals Along the *x*-Axis

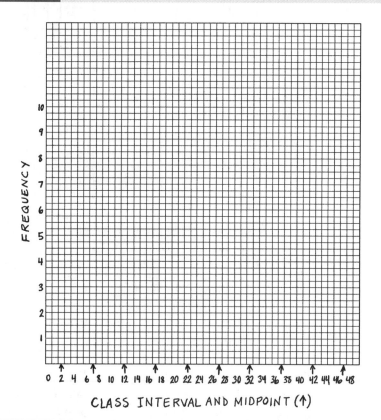

2. Draw a bar or column around each **midpoint** that represents the entire class interval to the height representing the frequency of that class interval. For example, in Figure 4.2, you can see our first entry where the class interval of 0–4 is represented by the frequency of 1 (representing the one time a value between 0 and 4 occurs). Continue drawing bars or columns until each of the frequencies for each of the class intervals is represented. Here's a nice hand-drawn (really!) histogram for the frequency distribution of the 50 scores that we have been working with so far.

Notice how each class interval is represented by a range of scores along the x-axis.

Figure 4.2 A Hand-Drawn Histogram

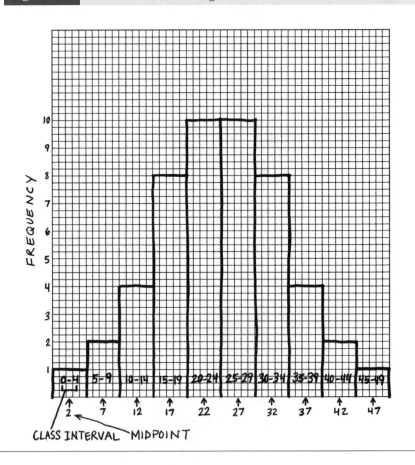

The Tally-Ho Method

You can see by the simple frequency distribution that you saw at the beginning of the chapter that you already know more about the distribution of scores than just a simple listing of them. You have a good idea of what values occur with what frequency. But another visual representation (besides a histogram) can be done by using tallies for each of the occurrences, as shown in Figure 4.3.

We used tallies that correspond with the frequency of scores that occur within a certain class. This gives you an even better visual representation of how often certain scores occur relative to other scores.

Figure 4.3	Tallying Scores

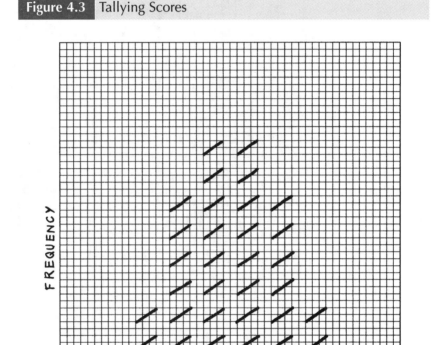

THE NEXT STEP: A FREQUENCY POLYGON

Creating a histogram or a tally of scores wasn't so difficult, and the next step (and the next way of illustrating data) is even easier. We're going to use the same data—and, in fact, the histogram that you just saw created—to create a frequency polygon. A **frequency polygon** is a continuous line that represents the frequencies of scores within a class interval, as shown in Figure 4.4.

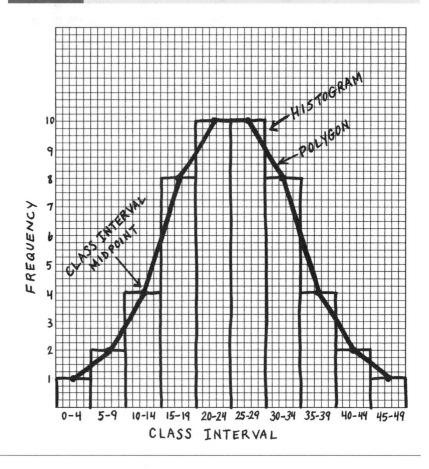

Figure 4.4   A Hand-Drawn Frequency Polygon

How did we draw this? Here's how.

1. Place a midpoint at the top of each bar or column in a histogram (see Figure 4.2).

2. Connect the lines and you've got it—a frequency polygon!

Note that in Figure 4.4, the histogram on which the frequency polygon is based is drawn using vertical and horizontal lines, and the polygon is drawn using curved lines. That's because, although we want you to see what a frequency polygon is based on, you usually don't see the underlying histogram.

Why use a frequency polygon rather than a histogram to represent data? It's more a matter of preference than anything else. A frequency polygon appears more dynamic than a histogram (a line that represents change in frequency always looks neat), but you are basically conveying the same information.

Cumulating Frequencies

Once you have created a frequency distribution and have visually represented those data using a histogram or a frequency polygon, another option is to create a visual representation of the cumulative frequency of occurrences by class intervals. This is called a **cumulative frequency distribution**.

A cumulative frequency distribution is based on the same data as a frequency distribution, but with an added column (Cumulative Frequency), as shown below.

Class Interval	Frequency	Cumulative Frequency
45–49	1	50
40–44	2	49
35–39	4	47
30–34	8	43
25–29	10	35
20–24	10	25
15–19	8	15
10–14	4	7
5–9	2	3
0–4	1	1

The cumulative frequency distribution begins by the creation of a new column labeled Cumulative Frequency. Then, we add the frequency in a class interval to all the frequencies below it. For example, for the class interval of 0–4, there is 1 occurrence and none below it, so the cumulative frequency is 1. For the class interval of 5–9, there are 2 occurrences in that class interval and one below it for a total of 3 (2 + 1) occurrences in that class interval or below it. The last class interval (45–49) contains 1 occurrence and there is a total of 50 occurrences at or below that class interval.

Once we create the cumulative frequency distribution, then the data can be plotted just as they were for a histogram or a frequency polygon. Only this time, we'll skip right ahead and plot the midpoint of each class interval as a function of the cumulative frequency of that class interval. You can see the cumulative frequency distribution in Figure 4.5 based on the 50 scores from the beginning of this chapter.

Figure 4.5	A Hand-Drawn Cumulative Frequency Distribution

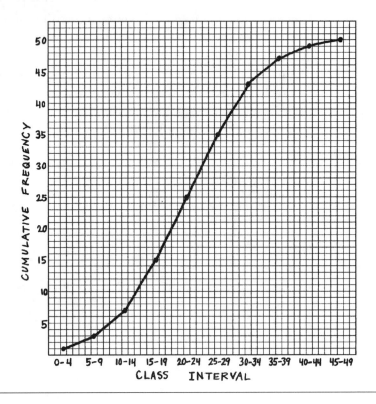

Another name for a cumulative frequency polygon is an **ogive.** And, if the distribution of the data is normal or bell shaped (see Chapter 8 for more on this), then the ogive represents what is popularly known as a bell curve or a normal distribution. SPSS creates a really nice ogive— it's called a P-P plot (for probability plot) and is really easy to create. See Appendix A for an introduction to creating graphs using SPSS, and also see the material toward the end of this chapter.

FAT AND SKINNY FREQUENCY DISTRIBUTIONS

You could certainly surmise by now that distributions can be very different from one another in a variety of ways. In fact, there are four different ways: average value, variability, skewness, and kurtosis. Those last two are new terms, and we'll define them as we show you what they look like. Let's define each of the four characteristics and then illustrate them.

Average Value

We're back once again to measures of central tendency. You can see in Figure 4.6 how three different distributions can differ in their average value. Notice that the average for Distribution C is more than the average for Distribution B, which, in turn, is more than the average for Distribution A.

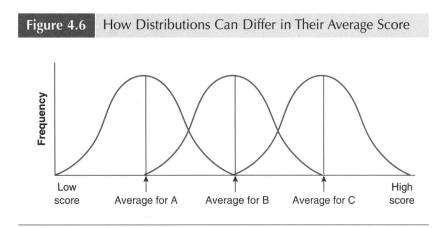

| Figure 4.6 | How Distributions Can Differ in Their Average Score |

Variability

In Figure 4.7, you can see three distributions that all have the same average value, but differ in variability. The variability in Distribution A is less than that in Distribution B, and, in turn, less than that found in C. Another way to say this is that Distribution C has the largest amount of variability of the three distributions, and A has the least.

Skewness

Skewness is a measure of the lack of symmetry, or the lopsidedness, of a distribution. In other words, one "tail" of the distribution is longer than another. For example, in Figure 4.8, Distribution A's right tail is longer than its left tail, corresponding to a smaller number of occurrences at the high end of the distribution. This is a positively skewed distribution. This might be the case when you have a test that is very difficult, and few people get scores that are relatively high and many more get scores that are relatively low. Distribution C's right tail is shorter than its left tail, corresponding to a larger number of occurrences at the high end of the distribution. This is a negatively skewed

Figure 4.7   How Distributions Can Differ in Variability

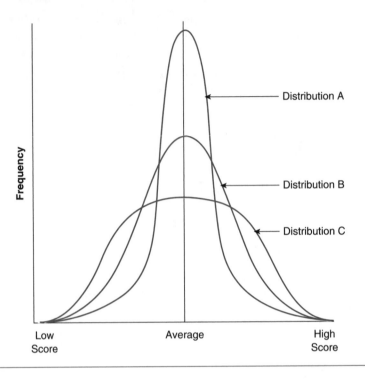

Figure 4.8   Degree of Skewness in Different Distributions

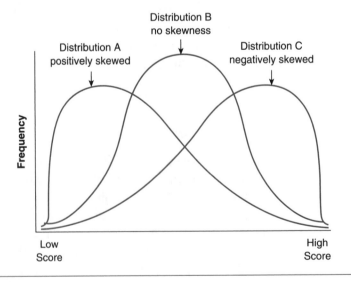

distribution and would be the case for an easy test (lots of high scores and relatively few low scores). And Distribution B—well, it's just right, equal lengths of tails and no skewness.

Kurtosis

Even though this sounds like a medical condition, it's the last of the four ways that we can classify how distributions differ from one another. **Kurtosis** has to do with how flat or peaked a distribution appears, and the terms used to describe this characteristic are relative ones. For example, the term **platykurtic** refers to a distribution that is relatively flat compared to a normal, or bell-shaped, distribution. The term **leptokurtic** refers to a distribution that is relatively peaked compared to a normal, or bell-shaped, distribution. In Figure 4.9, Distribution A is platykurtic compared to Distribution B. Distribution C is leptokurtic compared to Distribution B. Figure 4.9 looks similar to Figure 4.7 for a good reason—distributions that are platykurtic, for example, are relatively more disperse than those that are not. Similarly, a distribution that is leptokurtic is less variable or dispersed relative to others.

Figure 4.9	Degrees of Kurtosis in Different Distributions

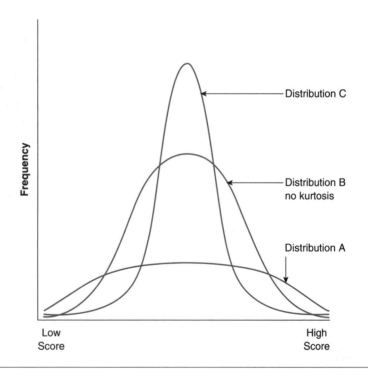

While skewness and kurtosis are used mostly as descriptive terms (such as "That distribution is negatively skewed"), there are mathematical indicators of how skewed or kurtotic a distribution is.

For example, skewness is computed by subtracting the value of the median from the mean. If the mean of a distribution is 100 and the median is 95, the skewness value is 100 – 95 = 5, and the distribution is positively skewed. If the mean of a distribution is 85 and the median is 90, the skewness value is 85 – 90 = –5, and the distribution is negatively skewed. There's an even more sophisticated formula, which is not relative, but takes the standard deviation of the distribution into account so that skewness indicators can be compared to one another (see Formula 4.1).

$$Sk = \frac{3(\overline{X} - M)}{s} \tag{4.1}$$

where

Sk is Pearson's (he's the correlation guy you'll learn about in Chapter 5) measure of skewness

$\overline{X}$ is the mean

M is the median

Here's an example: The mean of Distribution A is 100, the median is 105, and the standard deviation is 10. For Distribution B, the mean is 120, the median is 116, and the standard deviation is 10. Using Pearson's formula, the skewness of Distribution A is –1.5, and the skewness of Distribution B is 1.2. Distribution A is negatively skewed, and Distribution B is positively skewed. However, Distribution A is more skewed than Distribution B, regardless of the direction.

Let's not leave kurtosis out of this discussion. It too can be computed using a fancy formula as follows . . .

$$K = \frac{\sum \left(\frac{X - \overline{X}}{s} \right)^4}{n} - 3 \tag{4.2}$$

where

K = measure of kurtosis

Σ = sum

X = the individual score

$\overline{X}$ = the mean of the sample

s = the standard deviation

n = the sample size

This is a pretty complicated formula that basically looks at how flat or peaked a set of scores is. You can see that if each score is the same, then the numerator is zero and $K = 0$. K equals zero when the distribution is normal or mesokurtic (now there's a word to throw around). If the individual scores (the Xs in the formula) differ greatly from the mean (and there is lots of variability), then the curve will probably be quite peaked.

OTHER COOL WAYS TO CHART DATA

What we did so far in this chapter is take some data and show how charts such as histograms and polygons can be used to communicate visually. But there are several other types of charts that are used in the behavioral and social sciences, and although it's not necessary for you to know exactly how to create them (manually), you should at least be familiar with their names and what they do. So here are some popular charts, what they do, and how they do it.

There are several very good personal computer applications for creating charts, among them the spreadsheet Excel (a Microsoft product)—the author's personal favorite—and, of course, SPSS. For your information, the charts that you see in Figures 4.10, 4.11, and 4.12 were created using Excel. The charts in the "Using the Computer to Illustrate Data" section were created using SPSS.

Column Charts

A column chart should be used when you want to compare the frequencies of different categories with one another. Categories are organized horizontally on the x-axis, and values are shown vertically on the y-axis. Here are some examples of when you might want to use a column chart:

- Number of voters by political affiliation
- The sales of three different types of products
- Number of children in each of six different grades

Figure 4.10 shows a graph of number of voters by political affiliation.

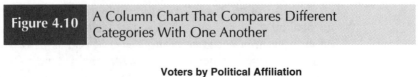

Figure 4.10 | A Column Chart That Compares Different Categories With One Another

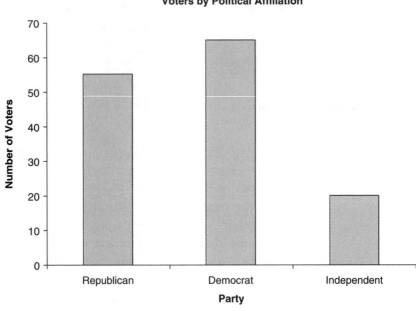

Bar Charts

A bar chart is identical to a column chart, but in this chart, categories are organized vertically on the y-axis and values are shown horizontally on the x-axis.

Line Charts

A line chart should be used when you want to show a trend in the data at equal intervals. Here are some examples of when you might want to use a line chart:

- Number of cases of mononucleosis (mono) per season among college students at three state universities
- Change in student enrollment over the school year
- Number of travelers on two different airlines for each quarter

In Figure 4.11, you can see a chart of the number of reported cases of mono by season among college students at three state universities.

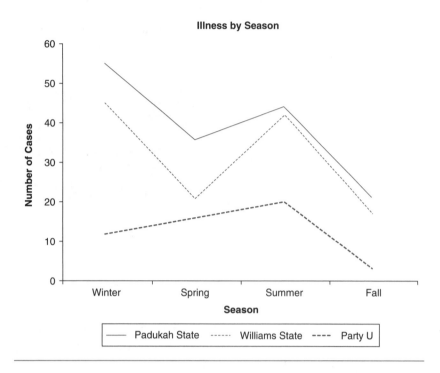

Figure 4.11 Using a Line Chart to Show a Trend Over Time

Illness by Season

Pie Charts

A pie chart should be used when you want to show the proportion of an item that makes up a series of data points. Here are some examples of when you might want to use a pie chart:

- Percentage of children living in poverty by ethnicity
- Proportion of night and day students enrolled
- Age of participants by gender

In Figure 4.12, you can see a pie chart of population by race. And, we did a few fancy-schmancy things such as making it 3-D and separating the slices.

Using the Computer (SPSS, That Is) to Illustrate Data

Now let's use SPSS and go through the steps in creating some of the charts that we explored in this chapter. First, some general SPSS charting guidelines.

Figure 4.12	A Pie Chart Illustrating the Relative Proportion of One Category to Others

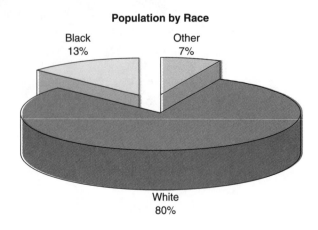

1. The newer versions of SPSS come with a Chart Builder option on the Graphs menu. This is the easiest way to get started, and that's what we will use.

2. In general, you click Graphs → Chart Builder, and you will see a dialog box from which you will select the type of graph you want to create.

3. Click the type of graph you want to create, and then select the specific design of that type of graph.

4. Drag the variable names to the axes where each belongs.

5. Click OK and you'll see your graph.

Let's practice.

Creating a Histogram Graph

1. Enter the data you want to use to create the graph. In this example, we will be using the same data that were used to create the histogram shown at the beginning of this chapter.

2. Click Graphs → Chart Builder and you will see the Chart Builder dialog box as shown in Figure 4.13. If you see any other screen, click OK.

3. Click the Histogram option in the Choose from: List and double-click the first image.

Figure 4.13   The Chart Builder Dialog Box

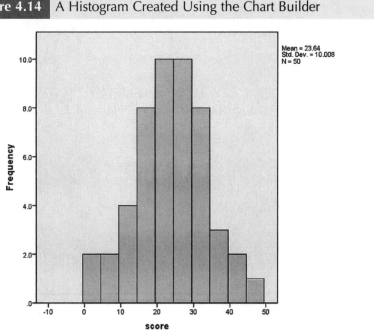

4. Drag the variable named Score to the *x*-axis? Location in the preview window.

5. Click OK and you will see the histogram as shown in Figure 4.14.

Figure 4.14   A Histogram Created Using the Chart Builder

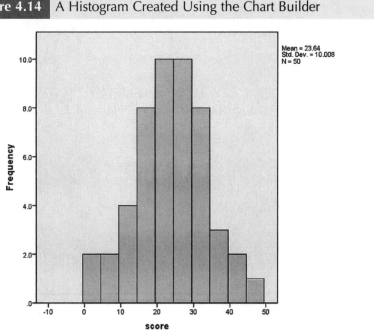

The histogram in Figure 4.14 looks a bit different from the one for 50 cases that was hand drawn and shown earlier in this chapter. The difference is that SPSS defines class intervals using its own idiosyncratic method. SPSS took as the middle of a class interval the bottom number of the interval (such as 10), rather than the midpoint (such as 12.5). Consequently, scores are allocated to different groups. The lesson here? How you group data makes a big difference in the way they look. And, once you get to know SPSS really well, you can make all kinds of fine-tuned adjustments to make graphs appear exactly as you want them.

Creating a Bar Graph

To create a bar graph, follow these steps.

1. Enter the data you want to use to create the graph. We used political party affiliation as you see here:

Republican	Democratic	Independent
54	63	19

2. Click Graphs → Chart Builder and you will see the Chart Builder dialog box as shown in Figure 4.13. If you see any other screen, click OK.

3. Click the Bar option in the Choose from: List and double-click the first image.

4. Drag the variable named Party to the x-axis? Location in the preview window.

5. Drag the variable named Count to the Count axis.

6. Click OK and you will see the bar graph as shown in Figure 4.15.

Figure 4.15 A Bar Graph Created Using the Chart Builder

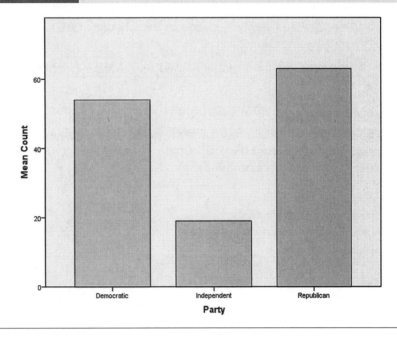

Creating a Line Graph

To create a line graph, follow these steps:

1. Enter the data you want to use to create the graph. In this example, we will be using the percentage in attendance of the total student body over the duration of a 10-year program. Here are the data:

Year	Percent Attending
1	87
2	88
3	89
4	76
5	80
6	96
7	91
8	97
9	89
10	79

2. Click Graphs → Chart Builder and you will see the Chart Builder dialog box as shown in Figure 4.13. If you see any other screen, click OK.

3. Click the Line option in the Choose from: List and double-click the first image.

4. Drag the variable named Year to the x-axis? Location in the preview window.

5. Drag the variable named Attendance to the y-axis? Location.

6. Click OK and you will see the line graph as shown in Figure 4.16. We used the SPSS Chart Editor to change the minimum and maximum values on the y-axis.

Figure 4.16   A Line Graph Created Using the Chart Builder

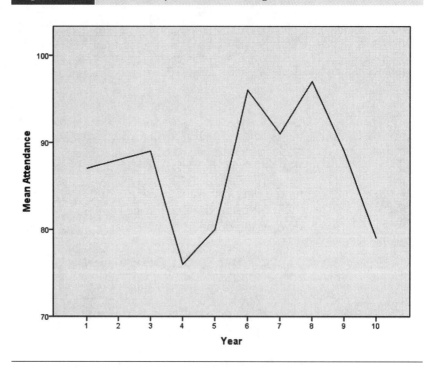

Creating a Pie Chart

To create a pie chart, follow these steps:

1. Enter the data you want to use to create the chart. In this example, the pie chart represents the percentage of people buying different brands of doughnuts. Here are the data:

Brand	Percentage
Krispies	55
Dunks	35
Other	10

2. Click Graphs → Chart Builder and you will see the Chart Builder dialog box as shown in Figure 4.13. If you see any other screen, click OK.

3. Click the Pie/Polar option in the Choose from: List and double-click the only image.

4. Drag the variable named Brand to the Slice by? axis label.

5. Drag the variable named Percentage to the Angle Variable? axis label.

6. Click OK and you will see the pie chart as shown in Figure 4.17.

Figure 4.17   A Pie Chart Created Using the Chart Builder

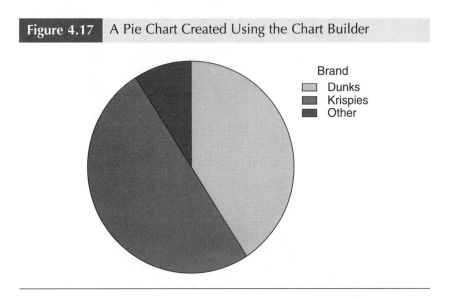

SUMMARY

There's no question that charts are fun to create and can add enormous understanding to what appears to be disorganized data. Follow our suggestions in this chapter and use charts well, but only when they enhance, not just add to, what's already there.

TIME TO PRACTICE

1. A data set of 50 comprehension scores (named Comprehension Score) called Chapter 4 Data Set 1 is available on the website. Answer the following questions and/or complete the following tasks:

   a. Create a frequency distribution and a histogram for the set.

   b. Why did you select the class interval you used?

   c. Is this distribution skewed? How do you know?

2. Here is a frequency distribution. Create a histogram either by hand or by using some other application such as Excel.

Class Interval	Frequency
90–100	12
80–89	14
70–79	20
60–69	24
50–59	28
40–49	29
30–39	21
20–29	15
10–19	17
0–9	12

3. A third-grade teacher is looking to improve her students' level of engagement during group discussions and instruction. She keeps track of each of the 15 third-graders' number of responses every day for 1 week, and it is available as Chapter 4 Data Set 2. Use SPSS to create a bar chart (one bar for each day—and warning, this may be a toughie).

4. Identify whether these distributions are negatively skewed, positively skewed, or not skewed at all, and why.

   a. This talented group of athletes scored very high on the vertical jump task.

   b. On this incredibly crummy test, everyone received the same score.

   c. On the most difficult spelling test of the year, the third graders wept as the scores were delivered.

5. For each of the following, indicate whether you would use a pie, line, or bar chart, and why.

   a. The proportion of freshmen, sophomores, juniors, and seniors in a particular university

   b. Change in GPA over four semesters

   c. Number of applicants for four different jobs

   d. Reaction time to different stimuli

   e. Number of scores in each of 10 categories

6. Provide an example of when you might use each of the following types of charts. For example, you would use a pie chart to show the proportion of children in Grades 1 through 6 who receive a reduced-price lunch. When you are done, draw the fictitious chart by hand.

   a. Line

   b. Bar

   c. Pie

7. Go to the library and find a journal article in your area of interest that contains empirical data, but does not contain any visual representation of them. Use the data to create a chart. Be sure to specify what type of chart you are creating, and why you chose the one you did. You can create the chart manually or using SPSS or Excel.

8. Create the worst-looking chart that you can, crowded with chart and font junk. There's nothing that has as lasting an impression as a *bad* example.

5 Ice Cream and Crime

Computing Correlation Coefficients

Difficulty Scale ☺☺ (moderately hard)

WHAT ARE CORRELATIONS ALL ABOUT?

Measures of central tendency and measures of variability are not the only descriptive statistics that we are interested in using to get a picture of what a set of scores looks like. You have already learned that knowing the values of the one most representative score (central tendency) and a measure of spread or dispersion (variability) is critical for describing the characteristics of a distribution.

However, sometimes we are as interested in the relationship between variables—or, to be more precise, how the value of one variable changes when the value of another variable changes. The way we express this interest is through the computation of a simple correlation coefficient.

A **correlation coefficient** is a numerical index that reflects the relationship between two variables. The value of this descriptive statistic ranges between -1 and $+1$. A correlation between two variables is sometimes referred to as a bivariate (for two variables) correlation. Even more specifically, the type of correlation that we will talk about in the majority of this chapter is called the **Pearson product-moment correlation,** named for its inventor, Karl Pearson.

Tech Talk

The Pearson correlation coefficient examines the relationship between two variables, but both of those variables are continuous in nature. In other words, they are variables that can assume any value along some underlying continuum, such as height, age, test score, or income. But there is a host of other variables that are not continuous. They're called discrete or categorical variables, like race (such as black and white), social class (such as high and low), and political affiliation (such as Democrat and Republican). You need to use other correlational techniques, such as the point-biserial correlation, in these cases. These topics are for a more advanced course, but you should know they are acceptable and very useful techniques. We mention them briefly later on in this chapter.

There are other types of correlation coefficients that measure the relationship between more than two variables, and we'll leave those for the next statistics course (which you are looking forward to already, right?).

Types of Correlation Coefficients: Flavor 1 and Flavor 2

A correlation reflects the dynamic quality of the relationship between variables. In doing so, it allows us to understand whether variables tend to move in the same or opposite directions when they change. If variables change in the same direction, the correlation is called a **direct correlation** or a positive correlation. If variables change in opposite directions, the correlation is called an **indirect correlation** or a negative correlation. Table 5.1 shows a summary of these relationships.

Now, keep in mind that the examples in the table reflect generalities. For example, regarding time to completion and the number of items correct on a test: In general, the less time that is taken on a test, the lower the score.

Such a conclusion is not rocket science, because the faster one goes, the more likely one is to make careless mistakes, such as not reading instructions correctly. But of course, there are people who can go very fast and do very well. And there are people who go very slow and don't do well at all. The point is that we are talking about the performance of a group of people on two different variables. We are computing the correlation between the two variables for the group, not for any one particular person.

Table 5.1	Types of Correlations and the Corresponding Relationship Between Variables			
What Happens to Variable *X*	**What Happens to Variable *Y***	**Type of Correlation**	**Value**	**Example**
X increases in value	*Y* increases in value	Direct or positive	Positive, ranging from .00 to +1.00	The more time you spend studying, the higher your test score will be.
X decreases in value	*Y* decreases in value	Direct or positive	Positive, ranging from .00 to +1.00	The less money you put in the bank, the less interest you will earn.
X increases in value	*Y* decreases in value	Indirect or negative	Negative, ranging from −1.00 to .00	The more you exercise, the less you will weigh.
X decreases in value	*Y* increases in value	Indirect or negative	Negative, ranging from −1.00 to .00	The less time you take to complete a test, the more you'll get wrong.

THINGS TO REMEMBER

There are several (easy but important) things to remember about the correlation coefficient:

- A correlation can range in value from −1 to +1.
- The absolute value of the coefficient reflects the strength of the correlation. So a correlation of −.70 is stronger than a correlation of +.50. One of the frequently made mistakes regarding correlation coefficients occurs when students assume that a direct or positive correlation is always stronger (i.e., "better") than an indirect or negative correlation because of the sign and nothing else.
- A correlation always reflects the situation where there are at least two data points (or variables) per case.
- Another easy mistake is to assign a value judgment to the sign of the correlation. Many students assume that a negative

relationship is not good and a positive one is good. That's why, instead of using the terms "negative" and "positive," the terms "indirect" and "direct" communicate meaning more clearly.

- The Pearson product-moment correlation coefficient is represented by the small letter r with a subscript representing the variables that are being correlated. For example,

r_{xy} is the correlation between variable X and variable Y

$r_{\text{weight-height}}$ is the correlation between weight and height

$r_{\text{SAT•GPA}}$ is the correlation between SAT score and grade point average (GPA)

Tech Talk

The correlation coefficient reflects the amount of variability that is shared between two variables and what they have in common. For example, you can expect an individual's height to be correlated with an individual's weight because they share many of the same characteristics, such as the individual's nutritional and medical history, general health, and genetics. However, if one variable does not change in value and therefore has nothing to share, then the correlation between the two variables is zero. For example, if you computed the correlation between age and number of years of school completed, and everyone was 25 years old, there would be no correlation between the two variables because there's literally nothing (any variability) about age available to share.

Likewise, if you constrain or restrict the range of one variable, the correlation between that variable and another variable is going to be less than if the range is not constrained. For example, if you correlate reading comprehension and grades in school for very high-achieving children, you'll find the correlation lower than if you computed the same correlation for children in general. That's because the reading comprehension score of very high-achieving students is quite high and much less variable than it would be for all children. The moral? When you are interested in the relationship between two variables, try to collect sufficiently diverse data—that way, you'll get the truest representative result.

COMPUTING A SIMPLE CORRELATION COEFFICIENT

The computational formula for the simple Pearson product-moment correlation coefficient between a variable labeled X and a variable labeled Y is shown in Formula 5.1.

$$r_{xy} = \frac{n\Sigma XY - n\Sigma X\Sigma Y}{\sqrt{[n\Sigma X^2 - (\Sigma X)^2][n\Sigma Y^2 - (\Sigma Y)^2]}} \qquad (5.1)$$

where

r_{xy} 　is the correlation coefficient between X and Y

n 　is the size of the sample

X 　is the individual's score on the X variable

Y 　is the individual's score on the Y variable

XY 　is the product of each X score times its corresponding Y score

X^2 　is the individual's X score, squared

Y^2 　is the individual's Y score, squared

Here are the data we will use in this example:

X	Y	X^2	Y^2	XY
2	3	4	9	6
4	2	16	4	8
5	6	25	36	30
6	5	36	25	30
4	3	16	9	12
7	6	49	36	42
8	5	64	25	40
5	4	25	16	20
6	4	36	16	24
7	5	49	25	35
Total, Sum, or Σ	54	43	320	201	247

Before we plug the numbers in, let's make sure you understand what each one represents.

ΣX, or the sum of all the X values, is 54

ΣY, or the sum of all the Y values, is 43

ΣX^2, or the sum of each X value squared, is 320

ΣY^2, or the sum of each Y value squared, is 201

ΣXY, or the sum of the products of X and Y, is 247

> It's easy to confuse the sum of a set of values squared and the sum of the squared values. The sum of a set of values squared is taking values such as 2 and 3, summing them (to be 5), and then squaring that (which is 25). The sum of the squared values is taking values such as 2 and 3; squaring them (to get 4 and 9, respectively); and then adding those together (to get 13). Just look for the parentheses as you work.

Here are the steps in computing the correlation coefficient:

1. List the two values for each participant. You should do this in a column format so as not to get confused.

2. Compute the sum of all the X values, and compute the sum of all the Y values.

3. Square each of the X values, and square each of the Y values.

4. Find the sum of the XY products.

These values are plugged into the equation you see in Formula 5.2:

$$r_{xy} = \frac{(10 \times 247) - (54 \times 43)}{\sqrt{[(10 \times 320) - 54^2][(10 \times 201) - 43^2]}} \tag{5.2}$$

Ta da! And you can see the answer in Formula 5.3:

$$r_{xy} = \frac{149}{213.83} = .692 \tag{5.3}$$

What's really interesting about correlations is that they measure the amount of distance that one variable co-varies in relation to another. So, if both variables are highly variable (have lots of wide-ranging scores), the correlation between them is more likely to be higher than if not. Now, that's not to say that lots of variability guarantees a higher correlation because the scores have to vary in a systematic way. But if the variance is contained in one variable, no matter how much the other variable changes, the correlation will be lower. For example, let's say you are examining the correlation between academic achievement in high school and first-year grades in college and you look at only the top 10% of the class. Well, that top 10% is likely to have very similar grades, introducing no variability and no room for the one variable to vary as a function of the other. Guess what you get when you correlate one variable with another that does not change? $r_{xy} = 0$, that's what. The lesson here? Variability works, and you should not artificially limit it.

A Visual Picture of a Correlation: The Scatterplot

There's a very simple way to visually represent a correlation: Create what is called a **scatterplot**, or **scattergram**. This is simply a plot of each set of scores on separate axes.

Here are the steps to complete a scattergram like you see in Figure 5.1 for the 10 sets of scores for which we computed the sample correlation above.

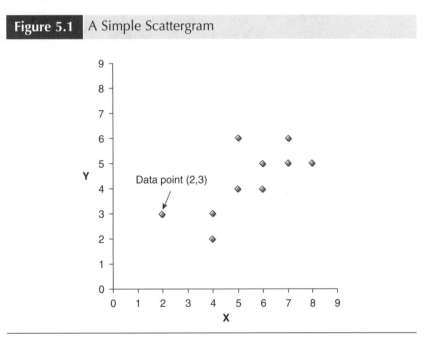

Figure 5.1   A Simple Scattergram

1. Draw the x-axis and the y-axis. Usually, the X variable goes on the horizontal axis and the Y variable goes on the vertical axis.

2. Mark both axes with the range of values that you know to be the case for the data. For example, the value of the X variable in our

example ranges from 2 to 8, so we marked the x-axis from 0 to 9. There's no harm in marking them a bit low or high—just as long as you allow room for the values to appear. The value of the *Y* variable ranges from 2 to 6, and we marked that axis from 0 to 9. Having similarly labeled axes can sometimes make the finished scatterplot easier to understand.

3. Finally, for each pair of scores (such as 2 and 3, as shown in Figure 5.1), we entered a dot on the chart by marking the place where 2 falls on the x-axis and 3 falls on the y-axis. The dot represents a data point, which is the intersection of the two values, as you can see in Figure 5.1.

When all the data points are plotted, what does such an illustration tell us about the relationship between the variables? To begin with, the general shape of the collection of data points indicates whether the correlation is direct (positive) or indirect (negative).

A positive slope occurs when the data points group themselves in a cluster from the lower left-hand corner on the x- and y-axes through the upper right-hand corner. A negative slope occurs when the data points group themselves in a cluster from the upper left-hand corner on the x- and y-axes through the lower right-hand corner.

Here are some scatterplots showing very different correlations where you can see how the grouping of the data points reflects the sign and strength of the correlation coefficient.

Figure 5.2 shows a perfect direct correlation where $r_{XY} = 1.00$ and all the data points are aligned along a straight line with a positive slope.

Figure 5.2 A Perfect Direct, or Positive Correlation

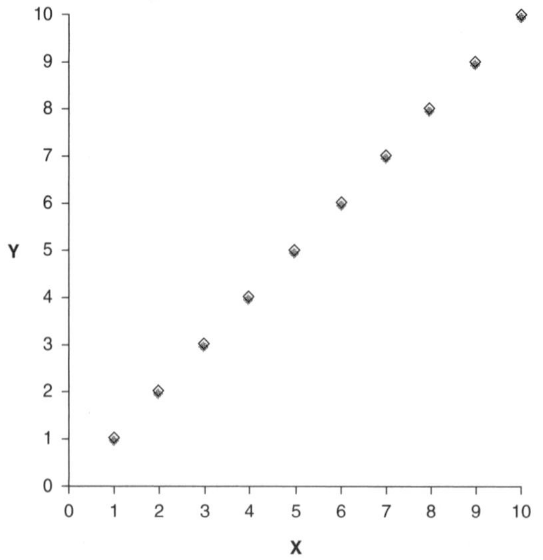

If the correlation were perfectly indirect, the value of the correlation coefficient would be –1.0 and the data points would align themselves in a straight line as well, but from the upper left-hand corner of the chart to the lower right. In other words, the line that connects the data points would have a negative slope.

> Don't ever expect to find a perfect correlation between any two variables in the behavioral or social sciences. It would say that two variables are so perfectly correlated, they share everything in common. In other words, knowing one is like knowing the other. Just think about your classmates. Do you think they all share any one thing in common that is perfectly related to another of their characteristics across all these different people? Probably not. In fact, *r* values approaching .7 and .8 are just about the highest you'll see.

In Figure 5.3, you can see the scatterplot for a strong (but not perfect) direct relationship where $r_{XY} = .70$. Notice that the data points align themselves along a positive slope, although not perfectly.

Now, we'll show you a strong indirect, or negative, relationship in Figure 5.4, where $r_{XY} = -.82$. Notice how the data points align themselves on a negative slope from the upper left-hand corner of the chart to the lower right-hand corner.

Figure 5.3   A Strong Positive, But Not Perfect, Direct Relationship

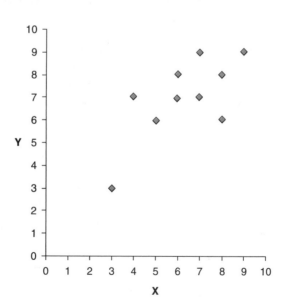

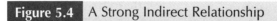

Figure 5.4 A Strong Indirect Relationship

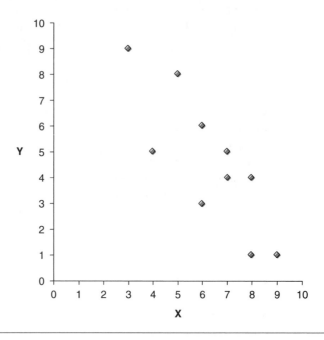

That's what different types of correlations look like, and you can really tell the general strength and direction by examining the way the points are grouped.

Tech Talk

Not all correlations are reflected by a straight line showing the X and the Y values in a relationship called a **linear correlation** (see Chapter 16 for tons of fun about this). The relationship may not be linear and may not be reflected by a straight line. Let's take the correlation between age and memory. For the early years, the correlation is probably highly positive—the older children get, the better their memories. Then, into young and middle adulthood, there isn't much of a change or much of a correlation, because most young and middle adults maintain a good memory. But with old age, memory begins to suffer, and there is an indirect relationship between memory and aging in the later years. If you take these together, you find that the correlation between memory and age tends to look something like a curve where memory increases, levels off, and then decreases. It's a curvilinear relationship, and sometimes, the best description of the relationship is a curvilinear one.

Bunches of Correlations: The Correlation Matrix

What happens if you have more than two variables? How are the correlations illustrated? Use a **correlation matrix** like the one shown below—a simple and elegant solution.

	Income	Education	Attitude	Vote
Income	1.00	0.574	−.08	−0.291
Educ	.574	1.00	−.149	−0.199
Attitude	−.08	−.149	1.00	−0.169
Vote	−.291	−.199	−.169	1.00

As you can see, there are four variables in the matrix: level of income (Income), level of education (Educ), attitude toward voting (Attitude), and whether the individual voted in the most recent election (Vote).

For each pair of variables, there is a correlation coefficient. For example, the correlation between income level and education is .574. Similarly, the correlation between income level and whether the person participated in the most recent election is −.291 (meaning that the higher the level of income, the less likely people are to vote).

In such a matrix with four variables, there are always $4!/(4 − 2)!2!$, or four things taken two at a time for a total of six correlation coefficients. Because variables correlate perfectly with themselves (those are the 1.00's down the diagonal), and because the correlation between Income and Vote is the same as the correlation between Vote and Income, the matrix creates a mirror image of itself.

You can use SPSS, and other statistical analysis packages such as Excel, to easily create a matrix like the one you saw earlier. You need to compute the bivariate values between a set of variables. In such applications as Excel, you can use the Data ToolPak.

You will see such matrices (the plural of matrix) when you read journal articles that use correlations to describe the relationship between several variables.

UNDERSTANDING WHAT THE CORRELATION COEFFICIENT MEANS

Well, we have this numerical index of the relationship between two variables, and we know that the higher the value of the correlation (regardless of its sign), the stronger the relationship is. But because the correlation coefficient is a value that is not directly tied to the value of an outcome, just how can we interpret it and make it a more meaningful indicator of a relationship?

Here are different ways to look at the interpretation of that simple r_{xy}.

Using-Your-Thumb Rule

Perhaps the easiest (but not the most informative) way to interpret the value of a correlation coefficient is by eyeballing it and using the information in Table 5.2.

So, if the correlation between two variables is .5, you could safely conclude that the relationship is a moderate one—not strong, but certainly not weak enough to say that the variables in question don't share anything in common.

This eyeball method is perfectly acceptable for a quick assessment of the strength of the relationship between variables, such as a description in a research report. But because this rule of thumb does depend on a subjective judgment (of what's "strong" or "weak"), we would like a more precise method. That's what we'll look at now.

Table 5.2   Interpreting a Correlation Coefficient

Size of the Correlation	Coefficient General Interpretation
.8 to 1.0	Very strong relationship
.6 to .8	Strong relationship
.4 to .6	Moderate relationship
.2 to .4	Weak relationship
.0 to .2	Weak or no relationship

A DETERMINED EFFORT: SQUARING THE CORRELATION COEFFICIENT

Here's the much more precise way to interpret the correlation coefficient: computing the coefficient of determination. The **coefficient of determination** is the percentage of variance in one variable that is accounted for by the variance in the other variable. Quite a mouthful, huh?

Earlier in this chapter, we pointed out how variables that share something in common tend to be correlated with one another. If we correlated math and English grades for 100 fifth-grade students, we would find the correlation to be moderately strong, because many of the reasons why children do well (or poorly) in math tend to be the same reasons why they do well (or poorly) in English. The number of hours they study, how bright they are, how interested their parents are in their schoolwork, the number of books they have at home, and more are all related to both math and English performance and account for differences between children (and that's where the variability comes in).

The more these two variables share in common, the more they will be related. These two variables share variability—or the reason why children differ from one another. And on the whole, the brighter child who studies more will do better.

To determine exactly how much of the variance in one variable can be accounted for by the variance in another variable, the coefficient of determination is computed by squaring the correlation coefficient.

For example, if the correlation between GPA and number of hours of study time is .70 (or $r_{GPA \bullet time} = .70$), then the coefficient of determination, represented by $r^2_{GPA \bullet time}$, is $.7^2$, or .49. This means that 49% of the variance in GPA can be explained by the variance in studying time. And the stronger the correlation, the more variance can be explained (which only makes good sense). The more two variables share in common (such as good study habits, knowledge of what's expected in class, and lack of fatigue), the more information about performance on one score can be explained by the other score.

However, if 49% of the variance can be explained, this means that 51% cannot—so even for a strong correlation of .70, many of the reasons why scores on these variables tend to be different from one another go unexplained. This amount of unexplained variance is called the **coefficient of alienation** (also called the **coefficient of nondetermination**). Don't worry. No aliens here. This isn't *X-Files* stuff, it's just the amount of variance in *Y* not explained by *X*.

How about a visual presentation of this sharing variance idea? OK. In Figure 5.5, you'll find a correlation coefficient, the corresponding coefficient of determination, and a diagram that represents how much variance is shared between the two variables. The larger the shaded area in each diagram (and the more variance the two variables share), the more highly the variables are correlated.

Figure 5.5	How Variables Share Variance and the Resulting Correlation

Correlation	Coefficient of Determination	Variable *X*		Variable *Y*
$r_{xy} = 0$	$r_{xy}^2 = 0$	◯	0% shared	◯
$r_{xy} = .5$	$r_{xy}^2 = .25$ or 25%		25% shared	
$r_{xy} = .9$	$r_{xy}^2 = .81$ or 81%		81% shared	

- The first diagram shows two circles that do not touch. They don't touch because they do not share anything in common. The correlation is 0.
- The second diagram shows two circles that overlap. With a correlation of .5 (and $r_{xy}^2 = .25$), they share about 25% of the variance between themselves.
- Finally, the third diagram shows that the two circles are almost placed one on top of the other. With an almost perfect correlation of $r_{xy} = .90$ ($r_{xy}^2 = .81$), they share about 81% of the variance between themselves.

As More Ice Cream Is Eaten . . . the Crime Rate Goes up (or Association vs. Causality)

Now here's the really important thing to be careful about when computing, reading about, or interpreting correlation coefficients.

Imagine this. In a small Midwestern town, a phenomenon was discovered that defied any logic. The local police chief observes that as ice cream consumption increases, crime rates tend to increase as well. Quite simply, if you measured both, you would find the relationship was direct, meaning that as people eat more ice cream, the crime rate increases. And as you might expect, as they eat less ice cream, the crime rate goes down. The police chief was baffled until he recalled the Stat 1 class he took in college and still fondly remembered.

His wondering how this could be turned into an aha! "Very easily," he thought. The two variables must share something or have something in common with one another. Remember that it must be something that relates to both level of ice cream consumption and level of crime rate. Can you guess what that is?

The *outside temperature* is what they both have in common. When it gets warm outside, such as in the summertime, more crimes are committed (it stays light longer, people leave the windows open, etc.). And because it is warmer, people enjoy the ancient treat and art of eating ice cream. And conversely, during the long and dark winter months, less ice cream is consumed and fewer crimes are committed as well.

Joe Bob, recently elected as a city commissioner, learns about these findings and has a great idea, or at least one that he thinks his constituents will love. (Keep in mind, he skipped the statistics offering in college.) Why not just limit the consumption of ice cream in the summer months, which will surely result in a decrease in the crime rate? Sounds good, right? Well, on closer inspection, it really makes no sense at all.

That's because of the simple principle that correlations express the association that exists between two or more variables; they have nothing to do with causality. In other words, just because level of ice cream consumption and crime rate increase together (and decrease together as well) does not mean that a change in one results in a change in the other.

For example, if we took all the ice cream out of all the stores in town and no more was available, do you think the crime rate would decrease? Of course not, and it's preposterous to think so. But strangely enough, that's often how associations are interpreted—as being causal in nature—and complex issues in the social and behavioral sciences are reduced to trivialities because of this misunderstanding. Did long hair and hippiedom have anything to do with the Vietnam conflict? Of course not. Does the rise in the number of crimes committed have anything to do with more efficient and safer cars? Of course not. But they all happen at the same time, creating the illusion of being associated.

OTHER COOL CORRELATIONS

There are different ways in which variables can be assessed. For example, nominal-level variables are categorical in nature, such as race (black or white) or political affiliation (Independent or Republican).

Or, if you are measuring income and age, these are both interval-level variables, because the underlying continuum on which they are based has equally appearing intervals. As you continue your studies, you're likely to come across correlations between data that occur at different levels of measurement. And to compute these correlations, you need some specialized techniques. Table 5.3 summarizes what these different techniques are and how they differ from one another.

Table 5.3   Correlation Coefficient Shopping, Anyone?

Level of Measurement and Examples			
Variable X	**Variable Y**	**Type of Correlation**	**Correlation Being Computed**
Nominal (voting preference, such as Republican or Democrat)	Nominal (sex, such as male or female)	Phi coefficient	The correlation between voting preference and sex
Nominal (social class, such as high, medium, or low)	Ordinal (rank in high school graduating class)	Rank biserial coefficient	The correlation between social class and rank in high school
Nominal (family configuration, such as intact or single parent)	Interval (grade point average)	Point biserial	The correlation between family configuration and grade point average
Ordinal (height converted to rank)	Ordinal (weight converted to rank)	Spearman rank coefficient	The correlation between height and weight
Interval (number of problems solved)	Interval (age in years)	Pearson correlation coefficient	The correlation between number of problems solved and age in years

USING THE COMPUTER TO COMPUTE A CORRELATION COEFFICIENT

Let's use SPSS to compute a correlation coefficient. The data set we are using is an SPSS data file named Chapter 5 Data Set 1.

There are two variables in this data set:

Variable	Definition
Income	Annual income in thousands of dollars
Education	Level of education measured in years

To compute the Pearson correlation coefficient, follow these steps:

1. Open the file named Chapter 5 Data Set 1.

2. Click Analyze → Correlate → Bivariate, and you will see the Bivariate Correlations dialog box as shown in Figure 5.6.

Figure 5.6   Identifying Variables for the Correlation Coefficient

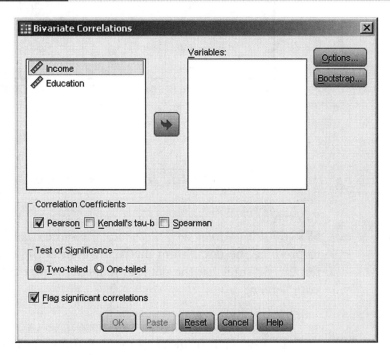

3. Double-click on the variable named Income to move it to the Variables: box.

4. Double-click on the variable named Educ to move it to the Variables: box. You can also hold down the Ctrl key to select more than one variable at a time.

5. Click OK.

The SPSS Output

The output in Figure 5.7 shows the correlation coefficient to be equal to .574. Also shown are the sample size 20 and a measure of the statistical significance of the correlation coefficient (we'll cover the topic of statistical significance in Chapter 7).

Figure 5.7	SPSS Output for the Computation of the Correlation Coefficient

Correlations

		Income	Education
Income	Pearson Correlation	1	.574**
	Sig. (2-tailed)		.008
	N	20	20
Education	Pearson Correlation	.574**	1
	Sig. (2-tailed)	.008	
	N	20	20

**. Correlation is significant at the 0.01 level (2-tailed).

Creating an SPSS Scatterplot (or Scattergram or Whatever)

You can do a scatterplot by hand, but it's good to know how to have SPSS do it for you as well. Let's take the same data that we just used to produce the correlation matrix in Figure 5.7, and create a scatterplot. Be sure that the data set named Chapter 5 Data Set 1 is on your screen.

1. Click Graphs → Chart Builder → Scatter/Dot and you will see the Chart Builder dialog box shown in Figure 5.8.

2. Double-click on the first example.

3. Drag the variable named Income to the *y*-axis?

4. Drag the variable named Education to the *x*-axis?

5. Click OK, and you'll have a very nice scatterplot like the kind you see in Figure 5.9.

Figure 5.8 The Chart Builder Dialog Box

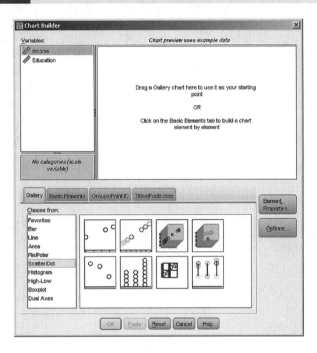

Figure 5.9 A Simple Scatterplot

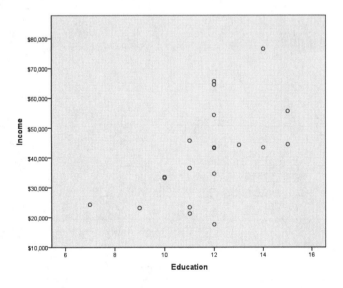

SUMMARY

The idea of showing how things are related to one another and what they have in common is a very powerful one and a very useful descriptive statistic (used in inference as well). Keep in mind that correlations express a relationship that is only associative and not causal, and you'll be able to understand how this statistic gives us valuable information about the relationships and how variables change or remain the same in concert with others. Now it's time to change speeds just a bit and wrap up Part I with a focus on reliability and validity. You need to know about these ideas because you'll be learning about looking at outcomes and what differences in scores and other variables represent.

TIME TO PRACTICE

1. Use these data to answer Questions 1a and 1b. These data are saved as Chapter 5 Data Set 2.

Total No. of Problems Correct (out of a possible 20)	Attitude Toward Test Taking (out of a possible 100)
17	94
13	73
12	59
15	80
16	93
14	85
16	66
16	79
18	77
19	91

a. Compute the Pearson product-moment correlation coefficient by hand and show all your work.

b. Construct a scatterplot for these 10 values by hand. Based on the scatterplot, would you predict the correlation to be direct or indirect? Why?

2. Use these data to answer Questions 2a and 2b.

Speed (to complete a 50-yard swim)	Strength (no. of pounds bench-pressed)
21.6	135
23.4	213
26.5	243
25.5	167
20.8	120
19.5	134
20.9	209
18.7	176
29.8	156
28.7	177

a. Using either a calculator or a computer, compute the Pearson correlation coefficient.

b. Interpret these data using the general range of very weak to very strong, and also compute the coefficient of determination. How does the subjective analysis compare to the value of r^2?

3. Rank the following correlation coefficients on strength of their relationship . . . (list the weakest first).

+.71

+.36

−.45

.47

−.62

4. For the following set of scores, calculate the Pearson correlation coefficient and interpret the outcome.

Achievement Increase	Classroom Budget Increase in 12 Months
7%	11%
3%	14%
5%	13%
7%	26%
2%	8%
1%	3%
5%	6%
4%	12%
4%	11%

5. For the following set of data, by hand, correlate minutes of exercise with GPA. What do you conclude given this analysis?

Exercise	GPA
25	3.6
30	4.0
20	3.8
60	3.0
45	3.7
90	3.9
60	3.5
0	2.8
15	3.0
10	2.5

6. Use SPSS to determine the correlation between hours of studying and grade point average in these honor students. Why is the correlation so low?

Hours of Studying	GPA
23	3.95
12	3.90
15	4.00
14	3.76
16	3.97
21	3.89
14	3.66
11	3.91
18	3.80
9	3.89

7. The coefficient of determination between two variables is .64. Answer the following questions:

a. What is the Pearson correlation coefficient?

b. How strong is the relationship?

c. How much of the variance in the relationship between these two variables is unaccounted for?

8. On the next page is a set of three variables (for each of 20 participants in a study on recovery from a head injury. Create a simple matrix that shows the correlations between each variable. You can do this by hand (and plan on being here for a while), or use SPSS or any other application.

Age at Injury	Level of Treatment	12-Month Treatment Score
25	1	78
16	2	66
8	2	78
23	3	89
31	4	87
19	4	90
15	4	98
31	5	76
21	1	56
26	1	72
24	5	84
25	5	87
36	4	69
45	4	87
16	4	88
23	1	92
31	2	97
53	2	69
11	3	79
33	2	69

9. Look at Table 5.3. What type of correlation coefficient would you use to examine the relationship between ethnicity (defined as different categories) and political affiliation? How about club membership (yes or no) and high school GPA? Explain why you selected the answers you did.

10. When two variables are correlated (such as strength and running speed), it also means that they are associated with one another. But if they are associated with one another, then why doesn't one cause the other?

6

Just the Truth

An Introduction to Understanding Reliability and Validity

Difficulty Scale ☺☺☺ (not so hard)

WHAT YOU'LL LEARN ABOUT IN THIS CHAPTER

✦ What reliability and validity are and why they are important

✦ This is a stat class! What's up with this measurement stuff?

✦ The basic measurement scales

✦ How to compute and interpret various types of reliability coefficients

✦ How to compute and interpret various types of validity coefficients

AN INTRODUCTION TO RELIABILITY AND VALIDITY

Professionals in the field of social welfare (and in other fields) recognize that the existence of more than a half million foster children in the United States is a serious concern. One of the major issues is how foster children adjust to their temporary adoptive families given that their biological families still play a very important role in their lives.

Sonya J. Leathers examined this question when she studied whether frequent parental visiting was associated with foster children's allegiances to foster families and biological parents. In a sample of 199 adolescents, she found that frequent visits to their family of origin did create conflicts, and she suggests interventions that might help minimize those conflicts.

To complete her study, she used a variety of different dependent variables (such as the Children's Symptom Inventory as well as interviews). Among other things, what she really got right is her care in selecting measurement instruments that had established, and acceptable, levels of reliability and validity—not a step that every researcher takes and one that we focus on in this chapter.

Want to know more? Check out the original reference: Leathers, S. (2003). Parental visiting, conflicting allegiances, and emotional and behavioral problems among foster children. *Family Relations, 52,* 53–63.

What's Up With This Measurement Stuff?

A very good question. After all, you enrolled in a stats class, and up to now, that's been the focus of the material that has been covered. Now it looks like you're faced with a topic that belongs in a tests and measurements class. So, what's this material doing in a stats book?

An excellent question, and one that you should be asking. Why? Well, much of what we have covered so far in *Statistics for People Who (Think They) Hate Statistics* has to do with the collection and description of data.

And, we are about to begin the journey toward analyzing and interpreting data. But before we begin learning those skills, we want to make sure that the data are what you think they are—that the data represent what it is you want to know about. In other words, if you're studying poverty, you want to make sure that the measure you use to assess poverty works. Or, if you are studying aggression in middle-aged males, you want to make sure that whatever tool you use to assess aggression works.

More really good news. . . . Should you continue in your education and want to take a class on tests and measurements, this introductory chapter will give you a real jump on understanding the scope of the area and what topics you'll be studying.

And in order to make sure that the entire process of collecting data and making sense out of them works, you first have to make sure that what you use to collect data works as well. The fundamental questions that will be answered in this chapter are "How do I know that the test, scale, instrument, etc. I use works every time I use it?" (that's reliability), and "How do I know that the test, scale, instrument, etc. I use measures what it is supposed to?" (that's validity).

Anyone who does research will tell you of the importance of establishing the reliability and validity of your test tool, whether it's a simple observational instrument of consumer behavior or one that measures a complex psychological construct such as attachment. However, there's another very good reason. If the tools that you use to collect data are unreliable or invalid, then the results of any test or any hypothesis have to be inconclusive. If you are not sure that the test does what it is supposed to and that it does so consistently, how do you know that the nonsignificant results you got are a function of the lousy test tools rather than an actual rejection of the null hypothesis when it is true (the Type I error friend you will learn about in the next chapter)? Want a clean test of the null? Make reliability and validity your business.

You may have noticed a new term at the beginning of this chapter—**dependent variable**. In an experiment, this is the outcome variable, or what the researcher looks at to see if any change has occurred as a function of the treatment that has taken place. And, guess what? The treatment has a name as well—the **independent variable**. For example, if a researcher examined the effect of different reading programs on comprehension, the independent variable would be the reading program and the dependent or outcome variable would be reading comprehension score. Although these terms will not be used often throughout the remainder of *Statistics for People . . .*, you should have some familiarity with them.

ALL ABOUT MEASUREMENT SCALES

Before we can talk much about reliability and validity, we first have to talk about different **scales of measurement**. What is measurement? Measurement is the assignment of values to outcomes following a set of rules—simple. The results are the different scales we'll define in a moment, and an outcome is anything we are interested in measuring, such as hair color, gender, test score, or height.

These scales of measurement, or rules, are particular levels at which outcomes are measured. Each level has a particular set of characteristics. And scales of measurement come in four flavors (there are four types): nominal, ordinal, interval, and ratio. Let's move on to a brief discussion and examples of the four scales of measurement.

A Rose by Any Other Name:
The Nominal Level of Measurement

The **nominal level of measurement** is defined by the characteristics of an outcome that fit into one and only one class or category. For example, gender can be a nominal variable (female and male); as can ethnicity (Caucasian or African American); as can political affiliation (Republican, Democrat, or Independent). Nominal-level variables are "names" (*nominal* in Latin), and this is the least precise level of measurement. Nominal levels of measurement have categories that are mutually exclusive; for example, political affiliation cannot be both Republican and Democrat.

Any Order Is Fine With Me:
The Ordinal Level of Measurement

The "ord" in **ordinal level of measurement** stands for order, and the characteristic of things being measured here is that they are ordered. The perfect example is a rank of candidates for a job. If we know that Russ is ranked #1, Sheldon is ranked #2, and Hannah is ranked #3, then this is an ordinal arrangement. We have no idea how much higher on this scale Russ is relative to Sheldon than Sheldon is relative to Hannah. We just know that it's better to be #1 than #2 than #3, but not by how much.

1 + 1 = 2: The Interval
Level of Measurement

Now we're getting somewhere. When we talk about the **interval level of measurement**, it is where a test or an assessment tool is based on some underlying continuum such that we can talk about how much more a higher performance is than a lesser one. For example, if you get 10 words correct on a vocabulary test, that is twice as many as getting five words correct. A distinguishing characteristic of interval-level scales is that the intervals along the scale are equal to one another. Ten words correct is two more than eight correct, which is three more than five correct.

Can Anyone Have Nothing of Anything?
The Ratio Level of Measurement

Well, here's a little conundrum for you. An assessment tool at the **ratio level of measurement** is characterized by the presence of an absolute zero on the scale. What that means is the absence of any of the trait that is being measured. The conundrum? Are there outcomes we measure where it is possible to have nothing of what is being measured? In some disciplines, that can be the case. For example, in the physical and biological sciences, you can have the absence of a characteristic, such as absolute zero (no molecular movement) or zero light. In the social and behavioral sciences, it's a bit harder. Even if you score 0 on that spelling test or miss every item of an IQ test (in Russian), it does not mean that you have no spelling ability or no intelligence, right?

In Sum . . .

These scales of measurement, or rules, represent particular levels at which outcomes are measured. And, in sum, we can say the following:

- Any outcome can be assigned to one of four scales of measurement.
- Scales of measurement have an order, from the least precise being nominal, to the most precise being ratio.
- The "higher up" the scale of measurement, the more precise the data being collected, and the more detailed and informative the data are. It may be enough to know that some people are rich and some poor (and that's a nominal or categorical distinction), but it's much better to know exactly how much money one makes (interval or ratio). We can always make the "rich"/"poor" distinction if we want to once we have all the information.
- Finally, the more precise scales (such as interval) contain all the qualities of the scales below them, including ordinal and nominal. If you know that the Bears' batting average is .350, you know it is better than the Tigers (who hit .250) by 100 points, but you also know that the Bears are better than the Tigers (but not by how much), and that the Bears are different from the Tigers (but there's no direction to the difference).

RELIABILITY-DOING IT AGAIN UNTIL YOU GET IT RIGHT

Reliability is pretty easy to figure out. It's simply whether a test, or whatever you use as a measurement tool, measures something consistently. If you administer a test of personality before a special treatment occurs, will the administration of that same test 4 months later be reliable? That, my friend, is one of the questions. And that is one reason why there are different types of reliability, each of which we will get to after we define reliability just a bit more.

Test Scores—Truth or Dare

When you take a test in this class, you get a score, such as 89 (good for you) or 65 (back to the books!). That test score consists of several different elements, including the **observed score** (or what you actually get on the test, such as 89 or 65) and a **true score** (the true, 100% accurate reflection of what you *really* know). We can't directly measure true score because it is a theoretical reflection of the actual amount of the trait or characteristic possessed by the individual.

Why aren't true scores and observed scores the same? Well, they can be if the test (and the accompanying observed score) is a perfect (and we mean absolutely perfect) reflection of what's being measured.

But the Yankees don't always win, the bread sometimes falls on the buttered side, and Murphy's Law tells us that the world is not perfect. So, what you see as an observed score may come close to the true score, but rarely are they the same. Rather, the difference as you see here is in the amount of error that is introduced.

Observed Score = True Score + Error Score

Error? Yes—in all its glory. For example, let's suppose for a moment that someone gets an 89 on a stats test, but his or her true score (which we never really know but only theorize about) is 80. That means that 9 points in the difference (that's the **error score**) are due to error, or the reason why individual test scores vary from being 100% true.

What might be the source of such error? Well, perhaps the room in which the test is taken is so warm that it causes you to fall asleep. That would certainly have an impact on your test score. Or, perhaps you didn't study for the test as much as you should have. Ditto. Both of these examples would reflect testing situations or conditions rather than qualities of the trait being measured, right?

Our job is to reduce those errors as much as possible by having, for example, good test-taking conditions and making sure you are encouraged to get enough sleep. Reduce the error and you increase the reliability, because the observed score more closely matches the true score.

The less error, the more reliable—it's that simple.

Different Types of Reliability

There are several different types of reliability, and we'll cover the four most important and most often used in this section. They are all summarized in Table 6.1.

Table 6.1	Different Types of Reliability, When They Are Used, How They Are Computed, and What They Mean

Type of Reliability	When You Use It	How You Do It	An Example of What You Can Say When You're Done
Test-retest reliability	When you want to know whether a test is reliable over time	Correlate the scores from a test given in Time 1 with the same test given in Time 2.	The Bonzo test of identity formation for adolescents is reliable over time.
Parallel forms reliability	When you want to know if several different forms of a test are reliable or equivalent	Correlate the scores from one form of the test with the scores from a second form of the same test of the same content (but not the exact same test).	The two forms of the Regular Guy test are equivalent to one another and have shown parallel forms reliability.

(Continued)

Table 6.1	(Continued)		
Type of Reliability	**When You Use It**	**How You Do It**	**An Example of What You Can Say When You're Done**
Internal consistency reliability	When you want to know if the items on a test assess one, and only one, dimension	Correlate each individual item score with the total score.	All of the items on the SMART Test of Creativity assess the same construct.
Interrater reliability	When you want to know whether there is consistency in the rating of some outcome	Examine the percent of agreement between raters.	The interrater reliability for the best-dressed Football player judging was .91, indicating a high degree of agreement between judges.

Test-Retest Reliability

Test-retest reliability is used when you want to examine whether a test is reliable over time.

For example, let's say that you are developing a test that will examine preferences for different types of vocational programs. You may administer the test in September and then readminster the same test (and it's important it be the same) again in June. Then, the two sets of scores (remember, the same people took it twice) are correlated and you have a measure of reliability. Test-retest reliability is a must when you are examining differences or changes over time.

You must be very confident that what you are measuring has been measured in a reliable way such that the results you are getting come as close as possible to the individual's score each and every time.

Computing Test-Retest Reliability. Here are some scores from a test at Time 1 and Time 2 for the MVE (Mastering Vocational Education) Test under development. Our goal is to compute the Pearson correlation coefficient as a measure of the test-retest reliability of the instrument.

ID	Scores From Time 1	Scores From Time 2
1	54	56
2	67	77
3	67	87
4	83	89
5	87	89
6	89	90
7	84	87
8	90	92
9	98	99
10	65	76

The first and last step in this process is to compute the Pearson product-moment correlation (see Chapter 5 for a refresher on this), which is equal to

$$r_{\text{Time1}\bullet\text{Time2}} = .90$$

What does .90 mean as far as test-retest reliability? We'll get to the interpretation of this value shortly.

Parallel Forms Reliability

Parallel forms reliability is used when you want to examine the equivalence or similarity between two different forms of the same test.

For example, let's say that you are doing a study on memory, and part of the task is to look at 10 different words, memorize them as best as you can, and then recite them back after 20 seconds of study and 10 seconds of rest. Because this is a study that takes place over a 2-day period and involves some training of memory skills, you want to have another set of items that is exactly similar in task demands, but obviously cannot be the same as far as content. So, you create another list of words that is hopefully similar to the first. In this example, you want the consistency to be high across forms— the same ideas being tested, just using a different form.

Computing Parallel Forms Reliability. Here are some scores from the IRMT (I Remember Memory Test) on Form A and Form B. Our goal is to compute the Pearson correlation coefficient as a measure of the parallel forms reliability of the instrument.

ID	Scores From Form A	Scores From Form B
1	4	5
2	5	6
3	3	5
4	6	6
5	7	7
6	5	6
7	6	7
8	4	8
9	3	7
10	3	7

The first and last step in this process is to compute the Pearson product-moment correlation (see Chapter 5 for a refresher on this), which is equal to

$$r_{\text{FormA•FormB}} = .13$$

We'll get to the interpretation of this value shortly.

Internal Consistency Reliability

Internal consistency reliability is quite different from the two previous types that we have explored. It is used when you want to know whether the items on a test are consistent with one another in that they represent one, and only one, dimension, construct, or area of interest.

Let's say that you are developing a test of attitudes toward different types of health care, and you want to make sure that the set of 5 items measures just that, and nothing else. You would look at the score for each item (for a group of test takers) and see if the individual score correlates with the total score. You would expect that people who scored high on certain items (e.g., "I like my HMO") would have scored low on others (e.g., "I don't like spending money on health care") and that this would be consistent across all the people who took the test.

Cronbach's alpha (or α) is a special measure of reliability known as internal consistency, where the more consistently individual item scores vary with the total score on the test, the higher the value. And, the higher the value, the more confidence you can have that this is a test that is internally consistent or measures one thing, and that one thing is the sum of what each item evaluates.

For example, here's a 5-item test that has lots of internal consistency. . . .

1. 4 + 4 = ?
2. 5 – ? = 3
3. 6 + 2 = ?
4. 8 – ? = 3
5. 1 + 1 = ?

All of the items seem to measure the same thing, regardless of what that same thing is (which is a validity question—stay tuned).

Now, here's a 5-item test that doesn't quite come up to speed as far as being internally consistent.

1. 4 + 4 = ?

2. Who is the fattest of the three little pigs?

3. 6 + 2 = ?

4. 8 – ? = 3

5. So, just what did the wolf want?

It's obvious why. These questions are inconsistent with one another—the key criterion for internal consistency.

Computing Cronbach's Alpha. Here are some sample data for 10 people on this 5-item attitude test (the I♥HMO Test) where scores are between 1 (*strongly disagree*) and 5 (*strongly agree*) on each item.

Tech Talk

When you compute Cronbach's alpha (named after Lee Cronbach), you are actually correlating the score for each item with the total score for each individual, and comparing that to the variability present for all individual item scores. The logic is that any individual test taker with a high total test score should have a high(er) score on each item (such as 5, 5, 3, 5, 3, 4, 4, 2, 4, 5) for a total score of 40, and that any individual test taker with a low(er) total test score should have a low(er) score on each individual item (such as 5, 1, 5, 1, 5, 5, 1, 5, 5, 1, 5, 1) for a total score of 40 as well, but much less unified or one-dimensional.

ID	Item 1	Item 2	Item 3	Item 4	Item 5
1	3	5	1	4	1
2	4	4	3	5	3
3	3	4	4	4	4
4	3	3	5	2	1
5	3	4	5	4	3
6	4	5	5	3	2
7	2	5	5	3	4
8	3	4	4	2	4
9	3	5	4	4	3
10	3	3	2	3	2

And, here's the formula to compute Cronbach's alpha:

$$\alpha = \left(\frac{k}{k-1}\right)\left(\frac{s_y^2 - \sum s_i^2}{s_y^2}\right) \tag{6.1}$$

where

k = the number of items

s_y^2 = the variance associated with the observed score

$\sum s_i^2$ = the sum of all the variances for each item

Here's the same set of data with the values (the variance associated with the observed score, or s_y^2, and the sum of all the variances for each item, or $\sum s_i^2$) needed to complete the above equation.

ID	Item 1	Item 2	Item 3	Item 4	Item 5	Total Score
1	3	5	1	4	1	14
2	4	4	3	5	3	19
3	3	4	4	4	4	19
4	3	3	5	2	1	14
5	3	4	5	4	3	19

ID	Item 1	Item 2	Item 3	Item 4	Item 5	Total Score
6	4	5	5	3	2	19
7	2	5	5	3	4	19
8	3	4	4	2	4	17
9	3	5	4	4	3	19
10	3	3	2	3	2	13
						$s_y^2 = 6.4$
Item Variance	0.32	0.62	1.96	0.93	1.34	$\sum s_i^2 = 5.17$

And when you plug all these figures into the equation and get the following,

$$\alpha = \left(\frac{5}{5-1}\right)\left(\frac{6.40 - 5.17}{6.4}\right) = .24 \qquad (6.2)$$

you find that coefficient alpha is .24 and you're done (except for the interpretation that comes later!).

If we told you that there were many other types of internal consistency validity, you would not be surprised, right? This is especially true for measures of internal consistency. Not only is there coefficient alpha, but also split-half reliability, Spearman-Brown, Kuder-Richardson 20 and 21 (KR_{20} and KR_{21}), and others that basically do the same thing—examine the one-dimensional nature of a test—only in different ways.

USING THE COMPUTER TO CALCULATE CRONBACH'S ALPHA

Once you know how to compute Cronbach's alpha by hand and want to move on to using SPSS, the transition is very easy. We are using the data set shown (starting on page 112, a 5-item test with 10 people's responses).

1. Enter the data in the Data Editor. Be sure that there is a separate column for each test item.

2. Click Analyze → Scale → Reliability Analysis, and you will see the dialog box shown in Figure 6.1.

The Reliability Analysis Dialog Box

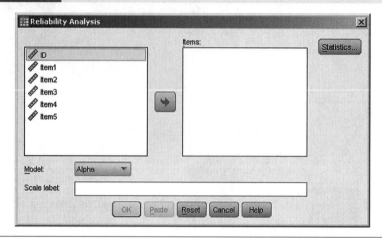

3. Move each of the variables (Item 1 through Item 5) to the Items box by double-clicking on it. Be sure, under Model, that Alpha is selected from the drop-down menu.

4. Click OK. SPSS will conduct the analysis and produce the output you see in Figure 6.2.

Figure 6.2   SPSS Output for the Reliability Analysis

Reliability

Scale: ALL VARIABLES

Case Processing Summary

		N	%
Cases	Valid	10	90.9
	Excluded[a]	1	9.1
	Total	11	100.0

a. Listwise deletion based on all variables in the procedure.

Reliability Statistics

Cronbach's Alpha	N of Items
.239	5

What the SPSS Output Means

As you can see, the value for alpha is .239, uncannily close to what we did by hand! The output does not tell you a whole lot more than this.

Interrater Reliability

Interrater reliability is the measure that tells you how much two raters agree on their judgments of some outcome.

For example, let's say you are interested in a particular type of social interaction during a transaction between a banker and a potential checking account customer, and you observe both people in real time (you're observing behind a one-way mirror) to see if the new and improved customer relations course that the banker took resulted in increased smiling and pleasant types of behavior toward the potential customer. Your job is to note every 10 seconds if the banker is demonstrating one of the three different behaviors he has been taught—smiling, leaning forward in his chair, or using his hands to make a point. Each time you see any one of those behaviors, you mark it on your scoring sheet as an X. If you observe nothing, you score a dash (–).

As part of this process, and to be sure that what you are recording is a reliable measure, you will want to find out what the level of agreement is between observers as to the occurrence of these behaviors. The more similar the ratings are, the higher the level of interrater agreement and interrater reliability.

Computing Interrater Reliability. In this example, the really important variable here is whether a customer-friendly act occurred within a set of 10-second time frames across 2 minutes (or twelve 10-second periods). So, what we are looking at is the rating consistency across a 2-minute time period broken down into twelve 10-second periods. An X on the scoring sheet means that the behavior occurred and a dash (–) means it did not.

	Time Period	1	2	3	4	5	6	7	8	9	10	11	12
Rater 1	Dave	X	–	X	X	X	–	X	X	–	–	X	X
Rater 2	Maureen	X	–	X	X	X	–	X	X	–	X	–	X

For a total of 12 periods (and 12 possible agreements), there are 7 where both Dave and Maureen agreed that agreement did take place (periods 1, 3, 4, 5, 7, 8, and 12), and 3 where they agreed it did not (periods 2, 6, and 9), for a total of 10 agreements and 2 disagreements.

Interrater reliability is computed using the following simple formula:

$$\text{Interrater reliability} = \frac{\text{number of agreements}}{\text{number of possible agreements}}$$

and when we plug in the numbers as you see here,

$$\text{Interrater reliability} = \frac{10}{12} = .833$$

the resulting interrater reliability coefficient is .833.

How Big Is Big? Finally: Interpreting Reliability Coefficients

OK, now we get down to business, and guess what? Remember all you learned about interpreting the value of the correlation coefficient in Chapter 5? It's almost the same as when you interpret reliability coefficients as well, with a (little) bit of a difference.

We want only two things, and here they are . . .

- Reliability coefficients to be positive and not to be negative
- Reliability coefficients that are as large as possible (between .00 and +1.00)

And If You Can't Establish Reliability . . . Then What?

The road to establishing the reliability of a test is not a smooth one at all, and not one that does not take a good deal of work. What if the test is not reliable?

Here are a few things to keep in mind. Remember that reliability is a function of how much error contributes to the observed score. Lower that error, and you increase the reliability.

- Make sure that the instructions are standardized and clear across all settings when the test is administered.
- Increase the number of items or observations, because the larger the sample from the universe of behaviors you are investigating, the more likely the sample is representative and reliable. This is especially true for achievement tests.
- Delete unclear items, because some people will respond in one way and others will respond in a different fashion, regardless of their knowledge or ability level or individual traits.
- For achievement tests especially (such as spelling or history tests), moderate the easiness and difficulty of tests, because any test that is too difficult or too easy does not reflect an accurate picture of one's performance.
- Minimize the effects of external events and standardize directions so that if a particularly important event, such as Mardi Gras or graduation, occurs near the time of testing, you can postpone any assessment.

Just One More Thing

The first step in creating an instrument that has sound psychometric (how's that for a big word?) properties is to establish its reliability (and we just spent some good time on that). Why? Well, if a test or measurement instrument is not reliable, is not consistent, and does not do the same thing time after time after time, it does not matter what it measures (and that's the validity question), right?

You could easily have the KACAS (Kids Are Cool at Spelling) test of introductory spelling and the first three items could be . . .

$16 + 12 = ?$

$21 + 13 = ?$

$41 + 33 = ?$

This is surely a highly reliable test, but surely not a valid one. Now that we have reliability well understood, let's move on to an introduction to validity.

VALIDITY—WHOA! WHAT IS THE TRUTH?

Validity is, most simply, the property of an assessment tool that indicates that the tool does what it says it does. A valid test is a test that measures what it is supposed to. If an achievement test is supposed to measure knowledge of history, then that's what it does. If an intelligence test is supposed to measure whatever intelligence is defined as by the test's creators, then it does just that.

Different Types of Validity

Just as there are different types of reliability, so there are different types of validity, and we'll cover the three most important categories and most often used in this section. They are all summarized in Table 6.2.

Table 6.2	Different Types of Validity, When They Are Used, How They Are Computed, and What They Mean

Type of Validity	When You Use It	How You Do It	An Example of What You Can Say When You're Done
Content validity	When you want to know whether a sample of items truly reflects an entire universe of items in a certain topic	Ask Mr. or Ms. Expert to make a judgment that the test items reflect the universe of items in the topic being measured.	My weekly quiz in my stat class fairly assesses the chapter's content.
Criterion validity	When you want to know if test scores are systematically related to other criteria that indicate that the test taker is competent in a certain area	Correlate the scores from the test with some other measure that is already valid and that assesses the same set of abilities.	The EATS test (of culinary skills) has been shown to be correlated with being a fine chef 2 years after culinary school (an example of predictive validity).
Construct validity	When you want to know if a test measures some underlying psychological construct	Correlate the set of test scores with some theorized outcome that reflects the construct for which the test is being designed.	It's true—men who participate in body contact and physically dangerous sports score higher on the TEST (osterone) test of aggression.

Content Validity

Content validity is the property of a test such that the test items sample the universe of items for which the test is designed. Content validity is most often used with achievement tests (e.g., everything from your first-grade spelling test to the Scholastic Aptitude Tests).

Establishing Content Validity. Establishing content validity is actually very easy. All you need is to locate your local (and cooperative) content expert. For example, if I were designing a test of introductory physics, I would go to the local physics expert (perhaps the teacher at the local high school or a professor at the university who teaches physics) and I would say, "Hey Albert (or Alberta), do you think this set of 100 multiple-choice questions accurately reflects all the possible topics and ideas that I would expect the students in my introductory class to understand?"

I would probably tell Albert or Alberta what the topics were and then he or she would look at the items and basically provide a judgment as to whether the items meet the criterion I established—a representation of the entire universe of all items that are introductory. If the answer is yes, I'm done (at least for now). If the answer is no, it's back to the drawing board and either the creation of new items or the refinement of existing ones.

Criterion Validity

Criterion validity assesses whether a test reflects a set of abilities in a current or future setting. If the criterion is taking place in the here and now, we talk about **concurrent criterion validity**. If the criterion is taking place in the future, we talk about **predictive validity**. For criterion validity to be present, one need not establish both concurrent and predictive validity, only the one that works for the purposes of the test.

Establishing Concurrent Validity. For example, you've been hired by the Universal Culinary Institute to design an instrument that measures culinary skills. Some part of culinary training has to do with straight knowledge (for example, what's a roux? And that's left to the achievement test side of things).

So, you develop a test that you think does a good job of measuring culinary skills, and now you want to establish the level of concurrent validity. To do this, you design the COOK scale, a set of 5-point items across a set of criteria (presentation, cleanliness, etc.) that each judge will use. As a criterion (and that's the key here), you have another set of judges rank each student from 1 to 10 on overall ability. Then, you simply correlate the COOK scores with the judge's rankings. If the validity coefficient (a simple correlation) is high, you're in business—if not, it's back to the drawing board.

Establishing Predictive Validity. Let's say that the cooking school has been percolating (heh-heh) along just fine for 10 years and you are interested not only in how well people cook (and that's the concurrent validity part of this exercise that you just established) but in the predictive validity as well. Now, the criterion changes from a here-and-now score (the one that judges give) to one that looks to the future.

Here, we are interested in developing a test that *predicts* success as a chef 10 years down the line. To establish the predictive validity of the COOK test, you go back and locate graduates of the program who have been out cooking for 10 years and administer the test to them. The criterion that is used here is their level of success, and you use as measures (a) whether they own their own restaurant, and (b) whether it has been in business for more than 1 year (given that the failure rate for new restaurants is more than 80% within the first year). The rationale here is that if a restaurant is in business for more than 1 year, then the chef must be doing something right.

To complete this exercise, you correlate the COOK score with a value of 1 (if the restaurant is in business for more than a year and owned by the graduate) with the previous (10 years earlier) COOK score. A high correlation coefficient indicates predictive validity, and a low correlation coefficient indicates the lack thereof.

Construct Validity

Construct validity is the most interesting and the most difficult of all the validities to establish because it is based on some underlying construct or idea behind a test or measurement tool.

You may remember from your extensive studies in Psych 1 that a construct is a group of interrelated variables. For example, aggression is a construct (consisting of such variables as inappropriate touching, violence, lack of successful social interaction, etc.), as is intelligence, mother-infant attachment, and hope. And keep in mind that these constructs are generated from some theoretical position that the researcher assumes. For example, he or she might propose that aggressive men have more trouble with the authorities than nonaggressive men.

Establishing Construct Validity. So, you have the FIGHT test (of aggression), which is an observational tool that consists of a series of items that is an outgrowth of your theoretical view about what the construct of aggression consists of. You know from the criminology literature that males who are aggressive do certain types of things more than others—for example, they get into more arguments, they are more physically aggressive (pushing and such), they commit more crimes of violence against others, and they have fewer successful interpersonal relationships. The FIGHT scale includes items that describe different behaviors, some of them theoretically related to aggressive behaviors and some that are not. Once the FIGHT scale is completed, you examine the results to see if positive scores on the FIGHT scale correlate with the presence of the kinds of behaviors you would predict (level of involvement in crime, quality of personal relationships, etc.) and don't correlate with the kinds of behaviors that should not be related (such as lack of domestic violence, completion of high school and college, etc.). And if the correlation is high for the items that you predict should correlate and low for the items that should not, then you can conclude that there is something about the FIGHT scale (and it is probably the items you designed that do not assess elements of aggression) that works. Congratulations.

And If You Can't Establish Validity . . . Then What?

Well, this is a tough one, especially because there are so many different types of validity.

In general, if you don't have the validity evidence you want, it's because your test is not doing what it should. If it's an achievement test, and a satisfactory level of content validity is what you seek, then you probably have to redo the questions on your test to make sure they are more consistent with what they should be according to that expert.

If you are concerned with criterion validity, then you probably need to reexamine the nature of the items on the test and answer the question of how well you would expect these responses to these questions to relate to the criterion you selected. And, of course, this assumes that the criterion you are using makes sense.

And finally, if it is construct validity that you are seeking and can't seem to find—better take a close look at the theoretical rationale that underlies the test you developed. Perhaps our definition and model of aggression is wrong, or perhaps intelligence needs some critical rethinking—all aimed at establishing consistency between the theory and the test items based on the theory.

A Last Friendly Word

This measurement stuff is pretty cool—intellectually interesting, and, in these times of accountability, everyone wants to know about the progress of students, stockbrokers, social welfare agency programs, and more.

Because of this strong and growing interest, there's a great temptation for undergraduate students working on their honors thesis or semester project or graduate students working on their thesis or dissertation to design an instrument for their final project.

But beware that what sounds like a good idea might lead to a disaster. The process of establishing the reliability and validity of any instrument can take years of intensive work. And what can make matters even worse is when the naïve or unsuspecting individual wants to create a new instrument to test a new hypothesis. That means that on top of everything else that comes with testing a new hypothesis, there is also the work of making sure the instrument works as it should.

> If you are doing original research of your own, such as for your thesis or dissertation requirement, be sure to find a measure that has already had reliability and validity evidence well established. That way, you can get on with the main task of testing your hypotheses and not fooling with the huge task of instrument development—a career in and of itself. Want a good start? Try the Buros Institute of Mental Measurements, available online at http://www.unl.edu/buros.

VALIDITY AND RELIABILITY: REALLY CLOSE COUSINS

Let's step back for a moment and recall one of the reasons that you're even reading this chapter.

It was assigned to you. No, really. This chapter is important because you need to know something about validity and the validity of the instruments you are using to measure outcomes. Why? If these instruments are not reliable *and* valid, then the results of your experiment will always be in doubt.

As we have mentioned earlier in this chapter, you can have a test that is reliable, but one that is not valid. However, you cannot have a valid test without it first being reliable. Why? Well, a test can do whatever it does over and over (that's reliability), but still not do what it is supposed to (that's validity). But if a test does what it is supposed to, then it has to do it consistently to work.

Tech Talk

You've read about the relationship between reliability and validity several places in this chapter, but there's a very cool relationship lurking out there that you may read about later in your coursework that you should know about now. This relationship says that the maximum level of validity is equal to the square root of the reliability coefficient. For example, if the validity coefficient for a test of mechanical aptitude is .87, the validity coefficient can be no larger than .93 (which is the square root of .87). What this means in tech talk is that the validity of a test is constrained by how reliable it is. And that makes perfect sense if we stop to think that a test must do what it does consistently before we are sure it does what it says it does.

But the relationship is closer as well. You cannot have a valid instrument without it first being reliable, because in order for something to do what it is supposed to do, it must first do it consistently, right? So, the two work hand in hand.

SUMMARY

Yep, this is a stats course, so what's the measurement stuff doing here? Once again, almost any use of statistics revolves around some outcome being measured. Just as you read basic stats to make sense of lots of data, you need basic measurement information to make sense out of how behaviors, test scores, and rankings and ratings are assessed.

TIME TO PRACTICE

1. Go to the library and find five journal articles in your area of interest in which reliability and validity data are reported, and discuss the outcome measures that are used. Identify the type of reliability that was established and the type of validity, and comment on whether you think that the levels are acceptable. If not, how can they be improved?

2. Provide an example of when you would want to establish test-retest and parallel forms reliability.

3. You are developing an instrument that measures vocational preferences (what you want to be), and you need to administer the test several times during the year while students are attending a vocational program. You need to assess the test-retest reliability of the test and the data from two administrations (available as Chapter 6

Data Set 1)—one in the Fall and one in the Spring. Would you call this a reliable test? Why or why not?

4. How can a test be reliable and not valid, and not valid unless it is reliable?

5. In general terms, describe what a test would be like if it were reliable but not valid. Now, do the same for a test that is valid, but not reliable.

6. When testing any experimental hypothesis, why is it important that the test you use to measure the outcome be both reliable and valid?

7. Describe the differences between content, criterion, and construct validity. Give examples of how each of these is measured.

PART III

Taking Chances for Fun and Profit

Snapshots

A scatterplot of student test scores.

W hat do you know so far, and what's next? To begin with, you've got a really solid basis for understanding how to describe the characteristics of a set of scores and how distributions can differ from one another. That's what you learned in Chapters 2, 3, and 4 of *Statistics for People Who (Think They) Hate Statistics*. In Chapter 5, you learned how to describe the relationship between variables using correlation tools. And, in Chapter 6, you learned about the importance of reliability and validity for understanding the integrity of any test score or other kind of outcome.

Now it's time to bump up the ante a bit and start playing for real. In Part III of *Statistics for People Who (Think They) Hate Statistics*, you will be introduced in Chapter 7 to the importance, and nature, of hypothesis testing, including an in-depth discussion of what a hypothesis is, what different types there are, the function of the hypothesis, and why and how hypotheses are tested.

Then, in Chapter 8, we'll get to the all-important topic of probability, represented by our discussion of the normal curve and the basic principles underlying probability—the part of statistics that helps us define how likely it is that some event (such as a specific score on a test) will occur. We'll use the normal curve as a basis for these arguments, and you'll see how any score or occurrence within any distribution has a likelihood associated with it.

After some fun with probability and the normal curve, we'll be ready to start our extended discussion in Part IV regarding the application of hypothesis testing and probability theory to the testing of specific questions regarding relationships between variables. It only gets better from here!

7 Hypotheticals and You

Testing Your Questions

Difficulty Scale ☺☺☺ (don't plan on going out tonight)

WHAT YOU'LL LEARN ABOUT IN THIS CHAPTER

✦ The difference between a sample and a population

✦ The importance of the null and research hypotheses

✦ The criteria for judging a good hypothesis

SO YOU WANT TO BE A SCIENTIST . . .

You might have heard the term *hypothesis* used in other classes. You may even have had to formulate one for a research project you did for another class, or you may have read one or two in a journal article. If so, then you probably have a good idea of what a hypothesis is. For those of you who are unfamiliar with this often-used term, a **hypothesis** is basically "an educated guess." Its most important role is to reflect the general problem statement or question that was the motivation for asking the research question in the first place.

That's why taking the care and time to formulate a really precise and clear research question is so important. This research question will be your guide in the creation of a hypothesis, and in turn, the hypothesis will determine the techniques you will use to test the hypothesis and answer the question that was originally asked.

So, a good hypothesis translates a problem statement or a research question into a form that is more amenable to testing. This form is called a hypothesis. We will talk about what makes a good hypothesis later in this chapter. Before that, let's turn our attention to the difference between a sample and a population. This is an important distinction because hypothesis testing deals with a sample, and then the results are generalized to the larger population. We'll then consider the two main categories of hypotheses (the null hypothesis and the research hypothesis). But first, let's formally define some simple terms that we have used earlier in *Statistics for People Who (Think They) Hate Statistics.*

Samples and Populations

As a good scientist, you would like to be able to say that if Method A is better than Method B, this is true forever and always and for all people in the universe, right? Indeed. And, if you do enough research on the relative merits of Methods A and B and test enough people, you may someday be able to say that. But don't get too excited, because it's unlikely you will be able to speak with such confidence. It takes too much money ($$$) and too much time (all those people!) to do all that research, and besides, it's not even necessary. Instead, you can just select a representative sample from the population and test your hypothesis about Methods A and B.

Given the constraints of never enough time and never enough research funds, with which almost all scientists live, the next best strategy is to take a portion of a larger group of participants and do the research with that smaller group. In this context, the larger group is referred to as a **population**, and the smaller group selected from that population is referred to as a **sample**.

A measure of how well a sample approximates the characteristics of a population is called **sampling error**. Sampling error is basically the difference between the values of the sample statistic and the population parameter. The higher the sampling error, the less precision one has in sampling and the more difficult it will be to make the case that what you find in the sample indeed reflects what you expect to find in the population.

Samples should be selected from populations in such a way that the sample matches as closely as possible the characteristics of the population. The goal is to have the sample as much like the population as possible. The most important implication of ensuring similarity between the two is that the research results based on the sample can be generalized to the population. When the sample accurately represents the population, the results of the study are said to have a high degree of generalizability.

A high degree of generalizability is an important quality of good research because it means that the time and effort (and $$$) that went into the research may have implications for groups of people other than the original participants.

THE NULL HYPOTHESIS

OK. So we have a sample of participants selected from a population, and to begin the test of our research hypothesis, we first formulate the **null hypothesis.**

The null hypothesis is an interesting little creature. If it could talk, it would say something like, "I represent no relationship between the variables that you are studying." In other words, null hypotheses are statements of equality demonstrated by the following real-life (brief) null hypotheses taken from a variety of popular social and behavioral science journals. Names have been changed to protect the innocent.

- There will be *no difference* in the average score of 9th graders and the average score of 12th graders on the ABC memory test.
- There is *no difference* between the effectiveness of community-based, long-term health care for older adults and the effectiveness of in-home, long-term health care on the social activities of older adults.
- There is *no relationship* between reaction time and problem-solving ability.
- There is *no difference* between white and black families in the amount of assistance offered to their children in school-related activities.

What these four null hypotheses have in common is that they all contain a statement that two or more things are equal, or unrelated, to each other.

The Purposes of the Null Hypothesis

What are the basic purposes of the null hypothesis? The null hypothesis acts as both a starting point and a benchmark against which the actual outcomes of a study can be measured.

Let's examine each of these purposes in more detail.

First, the null hypothesis acts as a starting point because it is the state of affairs that is accepted as true in the absence of any other information. For example, let's look at the first null hypothesis we stated above:

> There will be no difference in the average score of 9th graders and the average score of 12th graders on the ABC memory test.

Given absolutely no other knowledge of 9th and 12th graders' memory skills, you have no reason to believe that there will be differences between the two groups, right? If you know nothing about the relationship between these variables, the best you can do is guess. And that's taking a chance. You might speculate as to why one group might outperform another, but if you have no evidence a priori (before the fact), then what choice do you have but to assume that they are equal?

This lack of a relationship as a starting point is a hallmark of this whole topic. In other words, until you prove that there is a difference, you have to assume that there is no difference. And a statement of no difference or no relationship is exactly what the null hypothesis is all about.

Furthermore, if there are any differences between these two groups, you have to assume that these differences are due to the most attractive explanation for differences between any groups on any variable—chance! That's right: given no other information, chance is always the most likely and attractive explanation for the observed differences between two groups or the relationship between variables. Chance explains what we cannot. You might have thought of chance as the odds on winning that $5,000 nickel jackpot at the slots, but we're talking about chance as all that other "stuff" that clouds the picture and makes it even more difficult to understand the "true" nature of relationships between variables.

For example, you could take a group of soccer players and a group of football players and compare their running speeds. But look at all the factors we don't know about that could contribute to differences. Who is to know whether some soccer players practice more, or if some football players are stronger, or if both groups are receiving additional training?

What's more, perhaps the way their speed is being measured leaves room for chance; a faulty stopwatch or a windy day can contribute to differences unrelated to true running speed. As good researchers, our job is to eliminate chance factors from explaining observed differences and to evaluate other factors that might contribute to group differences, such as intentional training or nutrition programs, and see how they affect speed. The point is, if we find differences between groups and the differences are not due to training, we are at a loss as to what to attribute the difference to other than chance.

The second purpose of the null hypothesis is to provide a benchmark against which observed outcomes can be compared to see if these differences are due to some other factor. The null hypothesis helps to define a range within which any observed differences between groups can be attributed to chance (which is the null hypothesis's contention) or are due to something other than chance (which perhaps would be the result of the manipulation of some variable, such as training in the above example).

Most research studies have an implied null hypothesis, and you may not find it clearly stated in a research report or journal article. Instead, you'll find the research hypothesis clearly stated, which is now where we turn our attention.

THE RESEARCH HYPOTHESIS

Whereas a null hypothesis is a statement of no relationship between variables, a **research hypothesis** is a definite statement that there is a relationship between variables. For example, for each of the null hypotheses stated earlier, here is a corresponding research hypothesis. Notice that we said "a" and not "the" corresponding research hypothesis because there certainly could be more than one research hypothesis for any one null hypothesis.

- The average score of 9th graders *is different* from the average score of 12th graders on the ABC memory test.
- The effectiveness of community-based, long-term health care for older adults *is different* from the effectiveness of in-home, long-term health care on the social activities of older adults when measured using the Margolis Scale of Social Activities.
- Slower reaction time and problem-solving ability *are positively related.*
- There *is a difference* between white and black families in the amount of assistance offered to their children in educational activities.

Each of these four research hypotheses has one thing in common. They are all statements of inequality. They posit a relationship between variables and not an equality, as does the null hypothesis.

The nature of this inequality can take two different forms—a directional or a nondirectional research hypothesis. If the research hypothesis posits no direction to the inequality (such as "different from"), the hypothesis is a nondirectional research hypothesis. If the research hypothesis posits a direction to the inequality (such as "more than" or "less than"), the research hypothesis is a directional research hypothesis.

The Nondirectional Research Hypothesis

A **nondirectional research hypothesis** reflects a difference between groups, but the direction of the difference is not specified.
　For example, the research hypothesis

The average score of 9th graders is different from the average score of 12th graders on the ABC memory test

is nondirectional in that the direction of the difference between the two groups is not specified. The hypothesis states only that there is a difference and says nothing about the direction of that difference. It is a research hypothesis because a difference is hypothesized, but the nature of the difference is not specified.
　A nondirectional research hypothesis such as the one described here would be represented by the following equation:

$$H_1: \overline{X}_9 \neq \overline{X}_{12} \tag{7.1}$$

where

- H_1　represents the symbol for the first (of possibly several) research hypotheses
- $\overline{X}_9$　represents the average memory score for the sample of 9th graders
- $\overline{X}_{12}$　represents the average memory score for the sample of 12th graders
- $\neq$　means "is not equal to"

The Directional Research Hypothesis

A **directional research hypothesis** reflects a difference between groups, and the direction of the difference is specified.

For example, the research hypothesis

The average score of 12th graders is greater than the average score of 9th graders on the ABC memory test

is directional because the direction of the difference between the two groups is specified. One is hypothesized to be greater than (not just different from) the other.

An example of two other directional hypotheses is

A is greater than B (or A > B), or

B is greater than A (or A < B).

These both represent inequalities, but of a specific nature (greater than or less than). A directional research hypothesis such as the one described above, where 12th graders are hypothesized to score better than 9th graders, would be represented by the following equation:

$$H_1: \overline{X}_{12} > \overline{X}_9 \qquad\qquad (7.2)$$

where

H_1 represents the symbol for the first (of possibly several) research hypotheses

$\overline{X}_9$ represents the average memory score for the sample of 9th graders

$\overline{X}_{12}$ represents the average memory score for the sample of 12th graders

>   means "is greater than"

What is the purpose of the research hypothesis? It is this hypothesis that is directly tested as an important step in the research process. The results of this test are compared with what you expect by chance alone (reflecting the null hypothesis) to see which of the two is the more attractive explanation for any differences between groups you might observe.

In Table 7.1 are the four null hypotheses stated as both directional and nondirectional research hypotheses.

Table 7.1	Null Hypotheses and Corresponding Research Hypotheses	
Null Hypothesis	**Nondirectional Research Hypothesis**	**Directional Research Hypothesis**
There will be no difference in the average score of 9th graders and the average score of 12th graders on the ABC memory test.	Twelfth graders and 9th graders will differ on the ABC memory test.	Twelfth graders will have a higher average score on the ABC memory test than will 9th graders.
There is no difference between the effectiveness of community-based, long-term care for older adults and the effectiveness of in-home, long-term care on the Margolis Scale of Social Activities in older adults.	The effect of community-based, long-term care for older adults is *different* from the effect of in-home, long-term care on the social activities of older adults when measured using the Margolis Scale of Social Activities.	Older adults exposed to community-based, long-term care score higher on the Margolis Scale of Social Activities than do older adults receiving in-home, long-term care.
There is no relationship between reaction time and problem-solving ability.	There is a relationship between reaction time and problem-solving ability.	There is a positive relationship between reaction time and problem-solving ability.
There is no difference between white and black families in the amount of assistance offered to their children.	The amount of assistance offered by white families to their children is different from the amount of support offered by black families to their children.	The amount of assistance offered by white families to their children is more than the amount of support offered by black families to their children.

Tech Talk

Another way to talk about directional and nondirectional hypotheses is to talk about one- and two-tailed tests. A **one-tailed test** (reflecting a directional hypothesis) posits a difference in a particular direction, such as when we hypothesize that Group 1 will score higher than Group 2. A **two-tailed test** (reflecting a nondirectional hypothesis) posits a difference but in no particular direction. The importance of this distinction begins when you test different types of hypotheses (one- and two-tailed) and establish probability levels for rejecting or not rejecting the null hypothesis. More about this in Chapter 9. Promise.

Some Differences Between the Null
Hypothesis and the Research Hypothesis

Besides the null hypothesis representing an equality and the research hypothesis representing an inequality, there are several other important differences between the two types of hypotheses.

First, for a bit of review, the two types of hypotheses differ in that one (the null hypothesis) states that there is no relationship between variables (an equality), whereas the research hypothesis states that there is a relationship between the variables (an inequality). This is the primary difference.

Second, null hypotheses always refer to the population, whereas research hypotheses always refer to the sample. We select a sample of participants from a much larger population. We then try to generalize the results from the sample back to the population. If you remember your basic philosophy and logic (you did take these courses, right?), you'll remember that going from small (as in a sample) to large (as in a population) is a process of inference.

Third, because the entire population cannot be directly tested (again, it is impractical, uneconomical, and often impossible), you can't say with 100% certainty that there is no real difference between samples on some variable. Rather, you have to infer it (indirectly) from the results of the test of the research hypothesis, which is based on the sample. Hence, the null hypothesis must be indirectly tested, and the research hypothesis can be directly tested.

Fourth, null hypotheses are always written using Greek symbols, and research hypotheses are always written using Roman symbols. For example, the null hypothesis that the average score for 9th graders is equal to that of 12th graders is represented as you see here:

$$H_0: \mu_9 = \mu_{12} \qquad\qquad (7.3)$$

where

H_0 represents the null hypothesis

μ_9 represents the theoretical average for the population of 9th graders

μ_{12} represents the theoretical average for the population of 12th graders

The research hypothesis that the average score for a sample of 12th graders is greater than the average score for a sample of 9th graders is shown in Formula 7.2 (on page 133).

Finally, because you cannot directly test the null hypothesis, it is an *implied* hypothesis. But the research hypothesis is explicit and is stated as such. This is another reason why you rarely see null hypotheses stated in research reports and almost always see a statement of the research hypothesis.

WHAT MAKES A GOOD HYPOTHESIS?

You now know that hypotheses are educated guesses—a starting point for a lot more to come. As with any guess, some are better than others right from the start. We can't stress enough how important it is to ask the question you want answered and to keep in mind that any hypothesis you present is a direct extension of the original question you asked. This question will reflect your own personal interests and motivation and what research has been done previously. With that in mind, here are criteria you might use to decide whether a hypothesis you read in a research report or the ones you formulate are acceptable.

To illustrate, let's use an example of a study that examines the effects of after-school child care for employees who work late on the parents' adjustment to work. Here is a well-written hypothesis:

Parents who enroll their children in after-school programs will miss fewer days of work in one year and will have a more positive attitude toward work, as measured by the Attitude Toward Work survey, than will parents who do not enroll their children in such programs.

Here are the criteria.

First, a good hypothesis is stated in declarative form and not as a question. In the above example, the question, "Do you think parents and the companies they work for will be better . . . ?" was not posed because hypotheses are most effective when they make a clear and forceful statement.

Second, a good hypothesis posits an expected relationship between variables. The hypothesis that is being used as an example clearly describes the relationship between after-school child care, parents' attitude, and absentee rate. These variables are being tested to see if one (enrollment in the after-school program) has an effect upon the others (absentee rate and attitude).

Notice the word "expected" in the above criterion? Defining an expected relationship is intended to prevent the fishing trip (sometimes called the "shotgun" approach) that may be tempting to take but is not very productive.

Tech Talk

The fishing trip approach is where you throw out your line and take anything that bites. You collect data on as many things as you can, regardless of your interest or even whether collecting the data is a reasonable part of a scientific investigation. Or, you load up them guns and blast away at anything that moves, and you're bound to hit something. The problem is, you may not want what you hit, and, worse, you may miss what you want to hit, and worst of all (if possible), you may not know what you hit! Good researchers do not want just anything they can catch or shoot. They want specific results. To get them, researchers need their opening questions and hypotheses to be clear, forceful, and easily understood.

Third, hypotheses reflect the theory or literature on which they are based. As you read in Chapter 1, the accomplishments of scientists rarely can be attributed to just their own hard work. Their accomplishments are always due, in part, to many other researchers who came before them and laid the framework for later explorations. A good hypothesis reflects this, in that it has a substantive link to existing literature and theory. In the aforementioned example, let's assume that there is literature indicating that parents are more comfortable knowing their children are being cared for in a structured environment, and parents can then be more productive at work. Knowing this would allow one to hypothesize that an after-school program would provide the security parents are looking for. In turn, this allows them to concentrate on working rather than calling on the telephone to find out whether Rachel or Gregory got home safely.

Fourth, a hypothesis should be brief and to the point. You want your hypothesis to describe the relationship between variables in a declarative form and to be as direct and explicit as possible. The more to the point it is, the easier it will be for others (such as your master's thesis or doctoral dissertation committee members!) to read your research and understand exactly what you are hypothesizing and what the important variables are. In fact, when people read and evaluate research (as you will learn more about later in this chapter), the first thing many of them do is find the hypotheses to get a good idea as to the general purpose of the research and how things will be done. A good hypothesis tells you both of these things.

Fifth, good hypotheses are testable hypotheses. This means that you can actually carry out the intent of the question reflected by the hypothesis. You can see from the sample hypothesis that the important comparison is between parents who have enrolled their child in an after-school program and those who have not. Then, such things as attitude and work days missed will be measured. These are both reasonable objectives. Attitude is measured by the Attitude Toward Work survey (a fictitious title, but you get the idea), and absenteeism (the number of days away from work) is an easily recorded and unambiguous measure. Think how much harder things would be if the hypothesis were stated as *Parents who enroll their children in after-school care feel better about their jobs*. Although you might get the same message, the results might be more difficult to interpret given the ambiguous nature of words such as "feel better."

In sum, hypotheses should

- be stated in declarative form,
- posit a relationship between variables,
- reflect a theory or a body of literature on which they are based,
- be brief and to the point, and
- be testable.

When a hypothesis meets each of these five criteria, you know that it is good enough to continue with a study that will accurately test the general question from which the hypothesis was derived.

SUMMARY

A central component of any scientific study is the hypothesis, and the different types of hypotheses (null and research) help form a plan for answering the questions asked by the purpose of our research. The starting point and benchmark that characterize the null hypothesis let us use it as a comparison as we evaluate the acceptability of the research hypothesis. Now let's move on to how those null hypotheses are actually tested.

TIME TO PRACTICE

1. Go to the library and select five empirical (those containing data) research articles from your area of interest. For each one, list the following:

a. What is the null hypothesis (implied or explicitly stated)?

b. What is the research hypothesis (implied or explicitly stated)?

c. Using the content in any of these articles, create another research hypothesis.

2. Why does the scientific method work?

3. Why do good samples make for good tests of research hypotheses? What is the danger in not selecting a representative sample?

4. For the following research questions, create one null hypothesis, one directional research hypothesis, and one nondirectional research hypothesis.

a. What are the effects of attention on out-of-seat classroom behavior?

b. What is the relationship between the quality of a marriage and the quality of the spouses' relationships with their siblings?

c. What's the best way to treat an eating disorder?

5. Time to be even more creative! Provide one research hypothesis and an equation for each of the following topics:

a. The amount of money spent on food among undergraduate students and undergraduate student-athletes

b. The average amount of time taken by white and brown rats to get out of a maze

c. The effects of Drug A and Drug B on a disease

d. Method 1 and Method 2's time to complete a task

6. What do we mean when we say that the null hypothesis acts as a starting point?

7. Go back to the five hypotheses that you found in Question 1 above and evaluate each using the five criteria that were discussed at the end of the chapter.

8. Why does the null hypothesis presume no relationship between variables? (This is a tough one and a variation of Question 4 above.)

9. Create a research hypothesis tested using a one-tailed test and a research hypothesis tested using a two-tailed test.

8 Are Your Curves Normal?

Probability and Why It Counts

Difficulty Scale ☺☺☺ (not too easy and
not too hard, but very important)

WHAT YOU'LL LEARN ABOUT IN THIS CHAPTER

✦ Why understanding probability is basic to the understanding of statistics

✦ What the normal, or bell-shaped, curve is and what its characteristics are

✦ How to compute and interpret z scores

WHY PROBABILITY?

And here you thought this was a statistics class! Ha! Well, as you will learn in this chapter, the study of probability is the basis for the normal curve (much more on that later) and the foundation for inferential statistics.

Why? First, the normal curve provides us with a basis for understanding the probability associated with any possible outcome (such as the odds of getting a certain score on a test or the odds of getting a head on one flip of a coin).

Second, the study of probability is the basis for determining the degree of confidence we have in stating that a particular finding or outcome is "true." Or, better said, that an outcome (like an average score) may not have occurred due to chance alone. For example, let's compare Group A (which participates in 3 hours of extra swim practice each week) and Group B (which has no extra swim practice each week). We find that Group A differs from Group B on a test of fitness, but can we say that the difference is due to the extra practice or due to something else? The tools that the study of probability provides allow us to determine the exact mathematical likelihood that the difference is due to practice (and practice only) versus something else (such as chance).

All that time we spent on hypotheses in the previous chapter was time well spent. Once we put together our understanding of what a null hypothesis and a research hypothesis are with the ideas that are the foundation of probability, we'll be in a position to discuss how likely certain outcomes (formulated by the research hypothesis) are.

THE NORMAL CURVE (A.K.A. THE BELL-SHAPED CURVE)

What is a normal curve? Well, the **normal curve** (also called a **bell-shaped curve,** or bell curve) is a visual representation of a distribution of scores that has three characteristics. Each of these characteristics is illustrated in Figure 8.1.

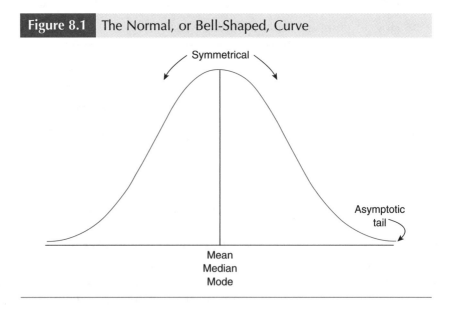

| Figure 8.1 | The Normal, or Bell-Shaped, Curve |

The normal curve represents a distribution of values where the mean, median, and mode are equal to one another. You probably remember from Chapter 4 that if the median and the mean are different, then the distribution is skewed in one direction or the other. The normal curve is not skewed. It's got a nice hump (only one), and that hump is right in the middle.

Second, the normal curve is perfectly symmetrical about the mean. If you fold one half of the curve along its center line, the two halves would fit perfectly on each other. They are identical. One-half of the curve is a mirror image of the other.

Finally (and get ready for a mouthful), the tails of the normal curve are **asymptotic**—a big word. What it means is that they come closer and closer to the horizontal axis, but never touch. See if you have some idea (in advance, because we will talk about it later) why this is so important because it's really a cornerstone of all this probability stuff.

The normal curve's shape of a bell also gives the graph its other name, the bell-shaped curve.

Tech Talk

When your devoted author was knee-high, he always wondered how the tail of a normal curve can approach the horizontal or *x*-axis yet never touch it. Try this. Place two pencils one inch apart and then move them closer (by half) so they are one-half inch apart, and then closer (one-quarter inch apart), and closer (one-eighth inch apart). They continually get closer, right? But they never (and never will) touch. Same thing with the tails of the curve. The tail slowly approaches the axis on which the curve "rests," but they can never really touch.

Why is this important? As you will learn later in this chapter, the fact that the tails never touch means that there is an infinitely small likelihood that a score can be obtained that is very extreme (way out in the left or right tail of the curve). If the tails did touch, then the likelihood that a very extreme score could be obtained would be nonexistent.

Hey, That's Not Normal!

We hope your next question is, "But there are plenty of sets of scores where the distribution is not normal or bell shaped, right?" Yes (and here comes the big *but*). When we deal with large sets of data (more than 30), and as we take repeated samples of the data from a population, the values in the curve closely approximate the shape of a normal curve. This is very important, because a lot of what we do when we talk about inferring from a sample to a population is based on the assumption that what is taken from a population is distributed normally.

And as it turns out, in nature in general, many things are distributed with the characteristics of what we call normal. That is, there are lots of events or occurrences right in the middle of the distribution, but relatively few on each end, as you can see in Figure 8.2, which shows the distribution of IQ and height in the general population.

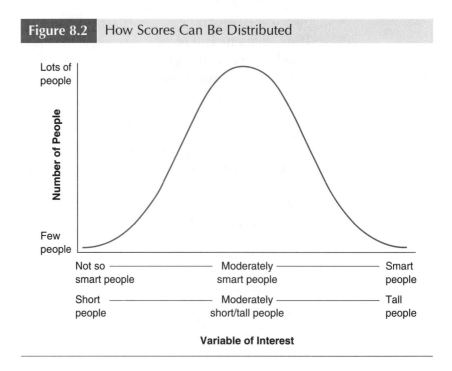

| Figure 8.2 | How Scores Can Be Distributed |

For example, there are very few people who are brilliant and very few who are intellectually or cognitively at the absolute bottom of the group. There are lots who are right in the middle and fewer as we move toward the tails of the curve. There are relatively few tall people and relatively few short people, but lots of people right in the middle. In both of these examples, the distribution of intellectual skills and height approximate a normal distribution.

Consequently, those events that tend to occur in the extremes of the normal curve have a smaller probability associated with each occurrence. We can say with a great deal of confidence that the odds of any one person (whose height we do not know beforehand) being very tall (or very short) are just not very great. But we know that the odds of any one person being average in height, or right around the middle, are pretty good. Those events that tend to occur in the middle of the normal curve have a higher probability of occurring than do those in the extreme.

More Normal Curve 101

You already know the three main characteristics that make a curve normal or make it appear bell shaped, but there's more to it than that. Take a look at the curve in Figure 8.3.

| Figure 8.3 | A Normal Curve Divided Into Different Sections |

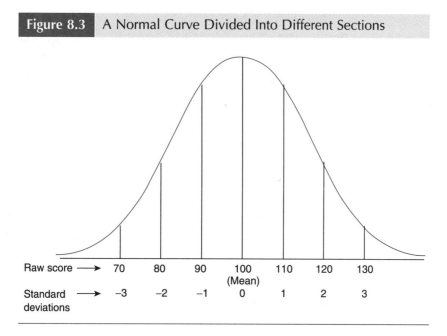

| Raw score → | 70 | 80 | 90 | 100 (Mean) | 110 | 120 | 130 |
| Standard deviations → | −3 | −2 | −1 | 0 | 1 | 2 | 3 |

The distribution represented here has a mean of 100 and a standard deviation of 10. We've added numbers across the x-axis that represent the distance in standard deviations from the mean for this distribution. You can see that the x-axis (representing the scores in the distribution) is marked from 70 through 130 in increments of 10 (which is the standard deviation for the distribution), the value of 1 standard deviation. We made up these numbers (100 and 10), so don't go nuts trying to find out where we got them from.

So, a quick review tells us that this distribution has a mean of 100 and a standard deviation of 10. Each vertical line within the curve separates the curve into a section, and each section is bound by particular scores. For example, the first section to the right of the mean of 100 is bound by the scores 100 and 110 representing 1 standard deviation from the mean (which is 100).

And below each raw score (70, 80, 90, 100, 110, 120, and 130), you'll find a corresponding standard deviation (−3, −2, −1, 0, +1, +2, and +3). As you may have figured out already, each standard deviation in our example is 10 points. So 1 standard deviation from the mean (which is 100) is the mean plus 10 points or 110. Not so hard, is it?

If we extend this argument further, then you should be able to see how the range of scores represented by a normal distribution with a mean of 100 and a standard deviation of 10 is 70 through 130 (which includes −3 to +3 standard deviations).

Now, here's a big fact that is always true about normal distributions, means, and standard deviations: For any distribution of scores (regardless of the value of the mean and standard deviation), if the scores are distributed normally, almost 100% of the scores will fit between –3 and +3 standard deviations from the mean. This is very important, because it applies to all normal distributions. Because the rule does apply (once again, regardless of the value of the mean or standard deviation), distributions can be compared with one another. We'll get to that again later.

With that said, we'll extend our argument a bit more. If the distribution of scores is normal, we can also say that between different points along the x-axis (such as between the mean and 1 standard deviation), a certain percentage of cases will fall. In fact, between the mean (which in this case is 100—got that yet?) and 1 standard deviation above the mean (which is 110), about 34% (actually 34.13%) of all cases in the distribution of scores will fall. This is a fact you can take to the bank because it will always be true.

Want to go further? Take a look at Figure 8.4. Here, you can see the same normal curve in all its glory (the mean equals 100 and the standard deviation equals 10)—and the percentage of cases that we would expect to fall within the boundaries defined by the mean and standard deviation.

Figure 8.4   A Normal Curve Divided Into Different Sections

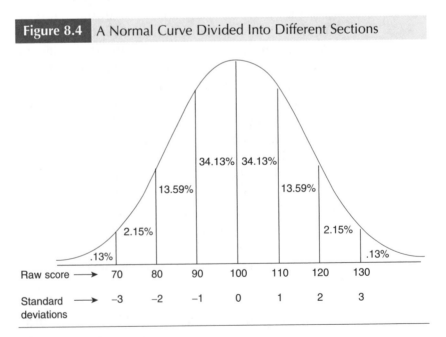

Here's what we can conclude.

The distance between	includes	and the scores that are included (if the mean = 100 and the standard deviation = 10) are from
The mean and 1 standard deviation	34.13% of all the cases under the curve	100 to 110
1 and 2 standard deviations	13.59% of all the cases under the curve	110 to 120
2 and 3 standard deviations	2.15% of all the cases under the curve	120 to 130
3 standard deviations and above	0.13% of all the cases under the curve	Above 130

If you add up all the values in either half of the normal curve, guess what you get? That's right, 50%. Why? The distance between the mean and all the scores to the right of the mean underneath the normal curve includes 50% of all the scores.

And because the curve is symmetrical about its central axis (each half is a mirror image of the other), the two halves together represent 100% of all the scores. Not rocket science, but important to point out, nonetheless.

Now, let's extend the same logic to the scores to the left of the mean of 100.

The distance between	includes	and the scores that are included (if the mean = 100 and the standard deviation = 10) are from
The mean and −1 standard deviation	34.13% of all the cases under the curve	90 to 100
−1 and −2 standard deviations	13.59% of all the cases under the curve	80 to 90
−2 and −3 standard deviations	2.15% of all the cases under the curve	70 to 80
−3 standard deviations and below	0.13% of all the cases under the curve	Below 70

Now, be sure to keep in mind that we are using a mean of 100 and a standard deviation of 10 only as sample figures for a particular example. Obviously, not all distributions have a mean of 100 and a standard deviation of 10.

All of this is pretty neat, especially when you consider that the values of 34.13% and 13.59% and so on are absolutely independent of the actual values of the mean and the standard deviation. This, roughly, is 34% because of the shape of the curve, not because of the value of any of the scores in the distribution or the value of the mean or standard deviation. In fact, if you actually drew a normal curve on a piece of cardboard and then cut out the area between the mean and +1 standard deviation and then weighed it, it would tip the scale at exactly 34.13% of the entire piece of cardboard from which the curve was cut. (Try it—it's true.)

In our example, this means that (roughly) 68% (34.13% doubled) of the scores fall between the raw score values of 90 and 110. What about the other 32%? Good question. One half (16%, or 13.59% + 2.15% + 0.13%) falls above (to the right of) 1 standard deviation above the mean and one half falls below (to the left of) 1 standard deviation below the mean. And because the curve slopes, and the amount of area decreases as you move farther away from the mean, it is no surprise that the likelihood that a score will fall more toward the extremes of the distribution is less than the likelihood it will fall toward the middle. That's why the curve has a bump in the middle and is not skewed in either direction.

OUR FAVORITE STANDARD SCORE: THE Z SCORE

You have read more than once how distributions differ in their central tendency and variability.

But in the general practice of research, we will find ourselves working with distributions that are indeed different, yet we will be required to compare them with one another. And to do such a comparison, we need some kind of a standard.

Say hello to **standard scores**. These are scores that are comparable because they are standardized in units of standard deviations. For example, a standard score of 1 in a distribution with a mean of 50 and a standard deviation of 10 means the same as a standard score of 1 from a distribution with a mean of 100 and a standard deviation of 5; they both represent 1 standard score and are an equivalent distance from their respective means. Also, we can use our knowledge of the normal curve and assign a probability to the occurrence of a value that is 1 standard deviation from the mean. We'll do that later.

Although there are other types of standard scores, the one that you will see most frequently in your study of statistics is called a **z score**. This is the result of dividing the amount that a raw score differs from the mean of the distribution by the standard deviation (see Formula 8.1):

$$z = \frac{(X - \overline{X})}{s}$$

(8.1)

where

z is the z score

X is the individual score

$\overline{X}$ is the mean of the distribution

s is the distribution standard deviation

For example, in Formula 8.2, you can see how the z score is calculated if the mean is 100, the raw score is 110, and the standard deviation is 10.

$$z = \frac{(110 - 100)}{10} = +1.0$$

(8.2)

It's just as easy to compute a raw score given a z score as the other way around. You already know the formula for a z score given the raw score, mean, and standard deviation. But if you know only the z score and the mean and standard deviation, then what's the corresponding raw score? Easy, just use the formula $X = z(s) + \overline{X}$. You can easily convert raw scores to z scores and back again if necessary. For example, a z score of –.5 in a distribution with a mean of 50 and an s of 5 would equal a raw score of $X = (-.5)(5) + 50$, or 47.5.

The following data show the original raw scores plus the z scores for a sample of 10 scores that has a mean of 12 and a standard deviation of 2. Any raw score above the mean will have a corresponding z score that is positive, and any raw score below the mean will have a corresponding z score that is negative. For example, a raw score of 15 has a corresponding z score of +1.5, and a raw score of 8 has a corresponding z score of −2.0. And of course, a raw score of 12 (or the mean) has a z score of 0 (which it must be because it is no distance from the mean).

X	$X - \bar{X}$	z Score
12	0	0
15	3	1.5
11	−1	−0.5
13	1	0.5
8	−4	−2
14	2	1
12	0	0
13	1	0.5
12	0	0
10	−2	−1

Below are just a few observations about these scores, as a little review.

First, those scores below the mean (such as 8 and 10) have negative z scores, and those scores above the mean (such as 13 and 14) have positive z scores.

Second, positive z scores always fall to the right of the mean and are in the upper half of the distribution. And negative z scores always fall to the left of the mean and are in the lower half of the distribution.

Third, when we talk about a score being located 1 standard deviation above the mean, it's the same as saying that the score is 1 z score above the mean. For our purposes, when comparing scores across distributions, z scores and standard deviations are equivalent. In other words, a z score is simply the number of standard deviations from the mean.

Finally (and this is very important), z scores across different distributions are comparable. Here's another table, similar to the one above, that will illustrate that last point. These 10 scores were selected from a set of 100 scores, with the scores having a mean of 59 and a standard deviation of 14.5.

Raw Score	$X - \bar{X}$	z Score
67	8	0.55
54	−5	−0.34
65	6	0.41
33	−26	−1.79
56	−3	−0.21
76	17	1.17
65	6	0.41
33	−26	−1.79
48	−11	−0.76
76	17	1.17

In the first distribution you saw earlier, with a mean of 12 and a standard deviation of 2, a raw score of 12.8 has a corresponding z score of +0.4, which means that a raw score of 12.8 is 0.4 standard deviations from the mean. In the second distribution, with a mean of 59 and a standard deviation of 14.5, a raw score of 64.8 has a corresponding z score of +0.4 as well. A miracle? No—just a good idea.

Both raw scores of 12.8 and 64.8, *relative to one another,* are equal distances (and equally distant) from the mean. When these raw scores are represented as standard scores, then they are directly comparable to one another in terms of their relative location in their respective distributions.

What z Scores Represent

You already know that a particular z score represents a raw score but also represents a particular location along the x-axis of a distribution. And the more extreme the z score (such as −2 or +2.6), the farther it is from the mean.

Because you already know the percentage of area that falls between certain points along the x-axis (such as 34% between the mean and a standard deviation of +1, for example, or about 14% between a standard deviation of +1 and a standard deviation of +2), we can make the following statements that will be true as well:

- 84% of all the scores fall below a z score of +1 (the 50% that falls below the mean plus the 34% that falls between the mean and the +1 z score).
- 16% of all the scores fall above a z score of +1 (because the total area under the curve has to equal 100%, and 84% of the scores fall below a score of +1.0).

Think about both of these for a moment. All we are saying is that, given the normal distribution, different areas of the curve are encompassed by different numbers of standard deviations or z scores.

OK—here it comes. These percentages or areas can also easily be seen as representing *probabilities* of a certain score occurring. For example, here's the big question (drum roll, please):

In a distribution with a mean of 100 and a standard deviation of 10, what is the probability that any one score will be 110 or above?

The answer? 16% or 16 out of 100 or .16. How did we get this?

First, we computed the corresponding z score, which is +1 [(110 – 100)/10]. Then, given the knowledge we already have (see Figure 8.4), we know a z score of 1 represents a location on the x-axis below which 84% (50% plus 34%) of all the scores in the distribution fall. Above that is 16% of the scores or a probability of .16. Because we already know the areas between the mean and 1, 2, or 3 standard deviations above or below the mean, we can easily figure out the probability that the value of any one z score has of occurring.

But the method we just went through is fine for z values of 1, 2, and 3. But what if the value of the z score is not a whole number like 2, but 1.23 or –2.01? We need to find a way to be more precise.

How do we do that? Simple—learn calculus and apply it to the curve to compute the area underneath it at almost every possible point along the x-axis, or (and we like this alternative much more) use Table B.1 found in Appendix B (the normal distribution table). This is a listing of all the values (except the very most extreme) for the area under a curve that corresponds to different z scores. This table has two columns. The first column, labeled *z Score,* is simply the z score that has been computed. The second column, *Area Between the Mean and the z Score,* is the exact area underneath the curve that is contained between the two points.

For example (and you should turn to page 351 and try this as you read along), if we wanted to know the area between the mean and a *z* score of +1, find the value 1.00 in the column labeled *z Score* and read across to the second column, where you find the area between the mean and a *z* score of 1.00 to be 34.13. Seen that before?

Why aren't there any plus or minus signs in this table (such as −1.00)? Because the curve is symmetrical, it does not matter if the value of the *z* score is positive or negative. The area between the mean and 1 standard deviation in any direction is always 34.13%.

Here's the next step. Let's say that for a particular *z* score of 1.38, you want to know the probability associated with that *z* score. If you wanted to know the percentage of the area between the mean and a *z* score of 1.38, you would find the corresponding area for the *z* score in Table B.1 of 1.38, which is 41.62, indicating that more than 41% of all the cases in the distribution fall within a *z* score of 0 and 1.38, and that about 92% (50% plus 41.62%) will fall at or below a *z* score of 1.38. Now, you should notice that we did this last example without any raw scores at all. Once you get to this table, they are just no longer needed.

But are we always interested only in the amount of area between the mean and some other *z* score? What about between two *z* scores, neither of which is the mean? For example, what if we were interested in knowing the amount of area between a *z* score of 1.5 and a *z* score of 2.5, which translates to a probability that a score falls between the two *z* scores? How can we use the table to compute these outcomes? It's easy. Just find the corresponding amount of area each *z* score encompasses and subtract one from the other. Often, drawing a picture helps, as in Figure 8.5.

For example, let's say that we want to find the area between a raw score of 110 and 125 in a distribution with a mean of 100 and a standard deviation of 10. Here are the steps we would take.

1. Compute the *z* score for a raw score of 110, which is (110 − 100)/10, or +1.

2. Compute the *z* score for a raw score of 125, which is (125 − 100)/10, or +2.5.

3. Using Table B.1 in Appendix B, find the area between the mean and a *z* score of +1, which is 34.13%.

4. Using Table B.1 in Appendix B, find the area between the mean and a *z* score of +2.5, which is 49.38%.

5. Because you want to know the distance between the two, subtract the smaller from the larger: 49.38 − 34.13, or 15.25%. Here's the picture that's worth a thousand words, in Figure 8.5.

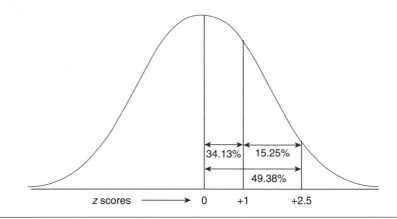

Figure 8.5 Using a Drawing to Figure Out the Difference in Area Between Two *z* Scores

OK—so we can be pretty confident that the probability of a particular score occurring can be best understood by examining where that score falls in a distribution relative to other scores. In this example, the probability of a score occurring between a *z* score of +1 and a *z* score of +2.5 is about 15%.

Here's another example. In a set of scores with a mean of 100 and a standard deviation of 10, a raw score of 117 has a corresponding *z* score of 1.70. This *z* score corresponds to an area under the curve of 95.54% (50% + 45.54%), meaning that the probability of this score occurring between a score of 0 and a score of 1.70 is 95.54% or 95.5 out of 100 or .955.

Tech Talk

Just two things about standard scores. First, even though we are focusing on *z* scores, there are other types of standard scores as well. For example, a *T* score is a type of standard score that is computed by multiplying the *z* score times 10 and adding 50. One advantage of this type of score is that you rarely have a negative *T* score. As with *z* scores, *T* scores allow you to compare standard scores from different distributions.

Second, a standard score is a whole different animal from a standardized score. A standardized score is one that comes from a distribution with a predefined mean and standard deviation. Standardized scores from tests such as the SATs and GREs (Graduate Record Exams) are used so that comparisons can easily be made between scores where the same mean and standard deviation are being used.

What z Scores Really Represent

The name of the statistics game is being able to estimate the probability of an outcome. If we take what we have talked about and done so far in this chapter one step further, it is deciding the probability of some event occurring. Then, we will use some criterion to judge whether we think that event is as likely, more likely, or less likely than what we would expect by chance. The research hypothesis presents a statement of the expected event, and we use our statistical tools to evaluate how likely that event is.

That's the 20-second version of what statistics is, but that's a lot. So let's take everything from this paragraph and go through it again with an example.

Let's say that your lifelong friend, trusty Lew, gives you a coin and asks you to determine if it is a "fair" one—that is, if you flip it 10 times you should come up with 5 heads and 5 tails. We would expect 5 heads (or 5 tails) because the probability is .5 of any one head or tail on any one flip. On 10 independent flips (meaning that one flip does not affect another), we should get 5 heads, and so on. Now the question is, how many heads would disqualify the coin as being fake or rigged?

Let's say the criterion for fairness that we will use is that if, in flipping the coin 10 times, we get heads (or heads turn up) less than 5% of the time, we'll say the coin is rigged and call the police on Lew (who, incidentally, is already on parole). This 5% criterion is a standard that is used by statisticians. If the probability of the event (be it the number of heads or the score on a test or the difference between the average scores for two groups) occurs in the extreme (and we're saying the extreme is defined as less than 5% of all such occurrences), it's an unlikely, or in this case, an unfair, outcome.

Here's the distribution of how many heads you can expect, just by chance alone on 10 flips. All the possible combinations are 2^{10} or 1,024 possible outcomes, such as 9 heads and 1 tail, 7 heads and 3 tails, and 10 heads and 0 tails, and on and on. For example, the probability associated with getting 6 heads in 10 flips is about 21%.

Number of Heads	Probability
0	0.00
1	0.01
2	0.04

(Continued)

(Continued)

Number of Heads	Probability
3	0.12
4	0.21
5	0.25
6	0.21
7	0.12
8	0.04
9	0.01
10	0.00

So, the likelihood of any particular outcome, such as 6 heads on 10 tosses, is about .21, or 21%. Now it's decision time. Just how many heads would one have to get on 10 flips to conclude that the coin is fixed, biased, busted, broken, or loony?

Well, as all good statisticians, we'll define the criterion as 5%, which we discussed. If the probability of the observed outcome (the results of all our flips) is less than 5%, we'll conclude that it is so unlikely that something other than chance must be responsible—and our conclusion will be that the "something other than chance" is a bogus coin.

If you look at the table, you can see that 8, 9, or 10 heads all represent outcomes that are less than 5%. So if the result of 10 coin flips was 8, 9, or 10 heads, the conclusion would be that the coin is not a fair one. (Yep—you're right, 0, 1, and 2 qualify for the same decision. Sort of the other side of the coin—groan.)

The same logic applies to our discussion of z scores earlier. Just how extreme a z score would we expect before we could proclaim that an outcome is not due just to chance but to some other factor? If you look at the normal curve table in Appendix B, you'll see that the cutoff point for a z score of 1.65 includes about 45% of the area under the curve. If you add that to the other 50% of the area on the other side of the curve, you come up with a total of 95%. That leaves just 5% above that point on the x-axis. Any score that represents a z score of 1.65 or above is then into pretty thin air—or at least in a location that has a much smaller chance of occurring than others.

Hypothesis Testing and z Scores: The First Step

What we showed you here is that any event can have a probability associated with it. And we use those probability values to make decisions as to how unlikely we think an event might be. For example, it's highly unlikely to get only 1 head and 9 tails in 10 tosses of a coin. And we also said that if an event seems to occur only 5 out of 100 times (5%), we will deem that event to be rather *unlikely* relative to all the other events that could occur.

It's much the same with any outcome related to a research hypothesis. The null hypothesis, which you learned about in Chapter 7, claims that there is no difference between groups (or variables), and that the likelihood of that occurring is 100%. We try to test the armor of the null for any chinks that might be there.

In other words, if, through the test of the research hypothesis, we find that the likelihood of an event that occurred is somewhat extreme, then the research hypothesis is a more attractive explanation than would be the null. So, if we find a z score (and remember that z scores have probabilities of occurrence associated with them as well) that is extreme (how extreme?—less than a 5% chance of occurring), we like to say that the reason for the extreme score is something to do with treatments or relationships and not just chance. We'll go into much greater detail on this point in the following chapter.

USING THE COMPUTER TO COMPUTE Z SCORES

SPSS does lots of really cool things, but it's the little treats like the one you'll see here that make the program such a great time saver. Now that you know how to compute z scores by hand, let's let SPSS do the work.

To have SPSS compute z scores for the set of data you see in the first column in Figure 8.6 (which you also saw earlier in the chapter on page 143), follow these steps.

1. Enter the data in a new SPSS window.

2. Click Analyze → Descriptive Statistics → Descriptives.

3. Double-click on the variable to move it to the Variable(s): box.

4. Click Save standardized values as variables in the Descriptives dialog box.

5. Click OK.

You can see in Figure 8.6 how SPSS data computes the corresponding z scores. (Be careful—when SPSS does almost anything, it automatically takes you to an Output window where you will not see the computed z scores! You have to switch back to the Data View.)

Figure 8.6   Having SPSS Compute z Scores for You

Score	ZScore
67	.62153
54	-.21145
65	.49338
33	-1.55703
56	-.08330
76	1.19821
65	.49338
33	-1.55703
48	-.59590
76	1.19821

SUMMARY

Being able to figure out a z score, and being able to estimate how likely it is to occur in a sample of data, is the first and most important skill in understanding the whole notion of inference. Once we know how likely a test score or a difference between groups is, we can then compare that likelihood to what we would expect by chance and then make informed decisions. As we start Part IV of *Statistics for People Who (Think They) Hate Statistics,* we'll apply this model to specific examples of testing questions about the difference.

TIME TO PRACTICE

1. What are the characteristics of the normal curve? What human behavior, trait, or characteristic can you think of that is distributed normally?

2. For a standard score, what three bits of information do you need?

3. Standard scores, such as z scores, allow us to make comparisons across different samples. Why?

4. Why is a z score a standard score, and why can standard scores be used to compare scores from different distributions with one another?

5. For the following set of scores, fill in the cells. The mean is 70 and the standard deviation is 8.

Raw Score	z Score
68.0	?
?	−1.6
82.0	?
?	1.8
69.0	?
?	−0.5
85.0	?
?	1.7
72.0	?

6. For the following set of scores, compute standard scores.
   18
   19
   15
   20
   25
   31
   17
   35
   27
   22
   34
   29
   40
   33
   21

7. Questions 7a through 7d are based on a distribution of scores with $\bar{X} = 75$ and the standard deviation = 6.38. Draw a small picture to help you see what's required.

   a. What is the probability of a score falling between a raw score of 70 and 80?

   b. What is the probability of a score falling above a raw score of 80?

    c. What is the probability of a score falling between a raw score of 81 and 83?

    d. What is the probability of a score falling below a raw score of 63?

8. Jake needs to score in the top 10% in order to earn a physical fitness certificate. The class mean is 78 and the standard deviation is 5.5. What raw score does he need to get that valuable piece of paper?

9. So, why doesn't it make sense to simply combine, for example, course grades across different topics—just take an average and call it a day?

10. Who is the better student, relative to his or her classmates? Here's all the information you ever wanted to know . . .

Math			
Class Mean	81		
Class Standard Deviation	2		
Reading			
Class Mean	87		
Class Standard Deviation	10		
Raw Scores			
	Math Score	Reading Score	Average
Noah	85	88	86.5
Talya	87	81	84
z Scores			
	Math Score	Reading Score	Average
Noah	_____	_____	_____
Talya	_____	_____	_____

PART IV

Significantly Different

Using Inferential Statistics

Snapshots

"I see that David here is already busy computing *z* scores..."

Y ou've gotten this far and you're still alive and kicking, so congratulate yourself. By this point, you should have a good understanding of what descriptive statistics is all about, how chance figures as a factor in making decisions about outcomes, and how likely outcomes are to have occurred due to chance or some treatment.

You're an expert on creating and understanding the role that hypotheses play in social and behavioral science research. Now it's time for the rubber to meet the road. Let's see what you're made of in the next part of *Statistics for People Who (Think They) Hate Statistics*. Best of all, the hard work you've put in will shortly pay off with an understanding of applied problems!

This part of the book deals exclusively with understanding and applying certain types of statistics to answer certain types of research questions. We'll cover the most common statistical tests, and even some that are a bit more sophisticated. At the end of this section, we'll show you some of the more useful software packages that can be used to compute the same values that we'll compute using a good old-fashioned calculator.

Let's start with a brief discussion of what the concept of significance is and go through the steps for performing an inferential test. Then we'll go on to examples of specific tests. We'll have lots of hands-on work here, so let's get started.

9

Significantly Significant

What It Means for You and Me

Difficulty Scale ☺☺ (somewhat
thought-provoking and key to it all!)

WHAT YOU'LL LEARN ABOUT IN THIS CHAPTER

✦ What the concept of significance is and why it is important

✦ The importance of and difference between Type I and Type II errors

✦ How inferential statistics work

✦ How to select the appropriate statistical test for your purposes

THE CONCEPT OF SIGNIFICANCE

There is probably no term or concept that represents more confusion for the beginning statistics student than the concept of statistical significance. But that doesn't mean it has to be that way for you. Although it's a powerful idea, it is also relatively straightforward and can be understood by anyone in a basic statistics class.

We need an example of a study to illustrate the points we want to make. Let's take E. Duckett and M. Richards's *Maternal Employment and Young Adolescents' Daily Experiences in Single-Mother Families* (paper presented at the Society for Research in Child Development, Kansas City, MO, 1989). These two authors examined the attitudes of 436 fifth- through ninth-grade adolescents toward maternal employment.

Specifically, the two researchers investigated whether differences are present between the attitudes of adolescents whose mothers work and the attitudes of adolescents whose mothers do not work.

They also examined some other factors, but for this example, we'll stick with the mothers-who-work and mothers-who-don't-work groups. One more thing. Let's add the word *significant* to our discussion of differences, so we have a research hypothesis something like this:

> There is a significant difference in attitude toward maternal employment between adolescents whose mothers work and adolescents whose mothers do not work, as measured by a test of emotional state.

What we mean by the word significant is that any difference between the attitudes of the two groups is due to some systematic influence and not due to chance. In this example, that influence is whether or not mothers work. We assume that all of the other factors that might account for any differences between groups were controlled. Thus, the only thing left to account for the differences between adolescents' attitudes is whether or not mothers work. Right? Yes. Finished? Not quite.

If Only We Were Perfect

Because our world is not a perfect one, we must allow for some leeway in how confident we are that only those factors we identify could cause any difference between groups. In other words, you need to be able to say that although you are pretty sure the difference between the two groups of adolescents is due to maternal employment, you cannot be absolutely, 100%, positively, unequivocally, indisputably (get the picture?) sure. There's always a chance, no matter how small, that you are wrong.

Why? Many reasons. For example, you could (horrors) just be plain ol' wrong. Maybe during this one experiment, differences between adolescents' attitudes were not due to whether mothers worked or didn't work, but were due to some other factor that was inadvertently not accounted for, such as a speech given by the local Mothers Who Work Club that several students attended. How about if the people in one group were mostly adolescent males and the people in the other group were mostly adolescent females? That could be the source of a difference as well. If you are a good researcher and do your homework, you can account for such differences, but it's always possible that you can't. And as a good researcher, you have to take that possibility into account.

So what do you do? In most scientific endeavors that involve testing hypotheses (such as the group differences example here), there is bound to be a certain amount of error that cannot be controlled—this is the chance factor that we have been talking about in the past few chapters. The level of chance or risk you are willing to take is expressed as a significance level, a term that unnecessarily strikes fear in the hearts of even strong men and women.

Significance level (here's the quick-and-dirty definition) is the risk associated with not being 100% confident that what you observe in an experiment is due to the treatment or what was being tested—in our example, whether or not mothers worked. If you read that significant findings occurred at the .05 level (or $p < .05$ in tech talk and what you regularly see in professional journals), the translation is that there is 1 chance in 20 (or .05 or 5%) that any differences found were not due to the hypothesized reason (whether mom works) but to some other, unknown reason or reasons (yep—or chance). Your job is to reduce this likelihood as much as possible by removing all of the competing reasons for any differences that you observed. Because you cannot fully eliminate the likelihood (because no one can control every potential factor), you assign some level of probability and report your results with that caveat.

In sum (and in practice), the researcher defines a level of risk that he or she is willing to take. If the results fall within the region that says, "This could not have occurred by chance alone—something else is going on," the researcher knows that the null hypothesis (which states an equality) is not the most attractive explanation for the observed outcomes. Instead, the research hypothesis (that there is an inequality or a difference) is the favored explanation.

Let's take a look at another example, this one hypothetical.

A researcher is interested in seeing whether there is a difference in the academic achievement of children who participated in a preschool program and children who did not participate. The null hypothesis is that the two groups are equal to each other on some measure of achievement.

The research hypothesis is that the mean score for the group of children that participated in the program is higher than the mean score for the group of children that did not participate in the program.

As a good researcher, your job is to show (as best you can—and no one is so perfect that he or she can account for everything) that *any* difference that exists between the two groups is due only to the effects of the preschool experience and no other factor or combination of factors. However, through a variety of techniques (that you'll learn about in your Stat II class!), you control or eliminate all the possible sources of difference, such as the influence of parents' education, number of children in the family, and so on. Once these other potential explanatory variables are removed, the only remaining alternative explanation for differences is the effect of the preschool experience itself.

But can you be absolutely (which is pretty darn) sure? No, you cannot. Why? First, because you can never be sure that you are testing a sample that identically reflects the profile of the population. And even if the sample perfectly represents the population, there are always other influences that might affect the outcome and that you inadvertently missed when designing the experiment. There's *always* the possibility of error.

By concluding that the differences in test scores are due to differences in treatment, you accept some risk. This degree of risk is, in effect (drumroll, please), the level of statistical significance at which you are willing to operate.

Statistical significance (here's the formal definition) is the degree of risk you are willing to take that you will reject a null hypothesis when it is actually true. For our example above, the null says that there is no difference between the two sample groups (remember, the null is always a statement of equality). In your data, however, you did find a difference. That is, given the evidence you have so far, group membership seems to have an effect on achievement scores. In reality, however, maybe there is no difference. If you reject the null you stated, you would be making an error. The risk you take in making this kind of error (or the level of significance) is also known as a Type I error.

The World's Most Important Table (for This Semester Only)

Here's what it all boils down to.

A null hypothesis can be true or false. Either there really is no difference between groups, or there really and truly is an inequality (such as the difference between two groups). But remember, you'll never know this true state because the null cannot be directly tested (remember that the null applies only to the population).

And, as a crackerjack statistician, you can choose to either reject or accept the null hypothesis. Right? These four different conditions create the table you see here in Table 9.1.

Let's look at each cell.

Table 9.1	Different Types of Errors

| | | Action You Take | |
		Accept the Null Hypothesis	**Reject the Null Hypothesis**
True nature of the null hypothesis	The null hypothesis is really true.	**1** ☐ Bingo, you accepted a null when it is true and there is really no difference between groups.	**2** Oops—you made a Type I error and rejected a null there really is no difference between groups. Type I errors are also represented by the Greek letter alpha, or a.
	The null hypothesis is really false.	**3** Uh-oh—you made a Type II error and accepted a false null hypothesis. Type II errors are also represented by the Greek letter beta, or β	**4** ☐ Good job, you rejected the null hypothesis when there really are differences between the two groups. This is also called power, or 1 – b.

More About Table 9.1

Table 9.1 has four important cells that describe the relationship between the nature of the null (whether it's true or not) and your action (accept or reject the null hypothesis). As you can see, the null can be either true or false, and you can either reject or accept it.

The most important thing about understanding this table is the fact that the researcher never really knows the true nature of the null hypothesis and whether there really is or is not a difference between groups. Why? Because the population (which the null represents) is never directly tested. Why? Because it's impractical to do so, and that's why we have inferential statistics.

- Cell 1 in Table 9.1 represents a situation where the null hypothesis is really true (there's no difference between groups) and the researcher made the correct decision in accepting it. No problem here. In our example, our results would show that there is no difference between the two groups of children, and we have acted correctly by accepting the null that there is no difference.

- Oops. Cell 2 represents a serious error. Here, we have rejected the null hypothesis (that there is no difference) when it is really true (and there is no difference). Even though there is no difference between the two groups of children, we will conclude there is and that's an error—clearly a boo-boo called a **Type I error**, also known as the level of significance.

- Uh-oh, another type of error. Cell 3 represents a serious error as well. Here, we have accepted the null hypothesis (that there is no difference) when it is really false (and, indeed, there is a difference). We have said that even though there is a difference between the two groups of children, we will conclude there is not—clearly a boo-boo, also known as a **Type II error.**

- Cell 4 in Table 9.1 represents a situation where the null hypothesis is really false and the researcher made the correct decision in rejecting it. No problem here. In our example, our results show that there is a difference between the two groups of children, and we have acted correctly by rejecting the null that states there is no difference.

So, if .05 is good and .01 is even better, why not set your Type I level of risk at .000001? For every good reason that you will be so rigorous in your rejection of false null hypotheses that you may miss a true one every now and then. Such a stringent Type I error rate allows for little leeway—indeed, the research hypothesis might be true but the associated probability might be .015—still quite rare, but missed with the too-rigid Type I level of error.

Back to Type I Errors

Let's focus a bit more on Cell 2, where a Type I error was made, because this is the focus of our discussion.

This Type I error, or level of significance, has certain values associated with it that define the risk you are willing to take in any test of the null hypothesis. The conventional levels set are between .01 and .05.

For example, if the level of significance is .01, it means that on any one test of the null hypothesis, there is a 1% chance you will reject the null hypothesis when the null is true and conclude that there is a group difference when there really is no group difference at all.

If the level of significance is .05, it means that on any one test of the null hypothesis, there is a 5% chance you will reject it when the null is true (and conclude that there is a group difference) when there really is no group difference at all. Notice that the level of significance is associated with an independent test of the null, and it is not appropriate to say that "on 100 tests of the null hypothesis, I will make an error on only 5, or 5% of the time."

In a research report, statistical significance is usually represented as $p < .05$, read as "the probability of observing that outcome is less than .05," often expressed in a report or journal article simply as "significant at the .05 level."

Tech Talk

With the introduction of fancy-schmancy statistical analysis software, there's no longer the worry about the imprecision of such statements as "$p < .05$" or "$p < .01$"—$p < .05$ can mean anything from .000 to .049999, right? Instead, software such as SPSS gives you the exact probability, such as $p = .013$ or $p = .158$, of the risk you are willing to take that you will commit a Type I error. So, when you see in a research article the statement that "$p < .05$," it means that the value of p is equal to anything from .00 to .049999999999 (you get the picture). Likewise, when you see "$p > .05$" or "$p = $ n.s." (for nonsignificant), it means that the probability of rejecting a true null exceeds .05 and, in fact, can range from .0500001 to 1.00.

So, it's actually terrific when we know the exact probability of an outcome because we can measure more precisely the risk we are willing to take.

There is another kind of error you can make, which, along with the Type I error, is shown in Table 9.1. A Type II error (Cell 3 in the chart) occurs when you inadvertently accept a false null hypothesis.

Tech Talk

When talking about the significance of a finding, you might hear the word *power* used. Power is a construct that has to do with how well a statistical test can detect and reject a null hypothesis when it is false. Mathematically, it's calculated by subtracting the value of the Type II error from 1. A more powerful test is always more desirable than a less powerful test, because the more powerful one lets you get to the heart of what's false and what's not.

For example, there may really be differences between the populations represented by the sample groups, but you mistakenly conclude there are not.

Ideally, you want to minimize both Type I and Type II errors, but it is not always easy or under your control. You have complete control over the Type I error level or the amount of risk that you are willing to take (because you actually set the level itself). Type II errors are not as directly controlled but, instead, are related to factors such as sample size. Type II errors are particularly sensitive to the number of subjects in a sample, and as that number increases, Type II error decreases. In other words, as the sample characteristics more closely match those of the population (achieved by increasing the sample size), the likelihood that you will accept a false null hypothesis decreases.

SIGNIFICANCE VERSUS MEANINGFULNESS

What an interesting situation for the researcher when discovering that the results of an experiment indeed are statistically significant. You know technically what statistical significance means—that the research was a technical success and the null hypothesis is not a reasonable explanation for what was observed. Now, if your experimental design and other considerations were well taken care of, statistically significant results are unquestionably the first step toward making a contribution to the literature in your field. However, the value of statistical significance and its importance or meaningfulness must be kept in perspective.

For example, let's take the case where a very large sample of illiterate adults (say, 10,000) is divided into two groups. One group receives intensive training to read using computers, and the other receives intensive training to read using classroom teaching. The average score for Group 1 (which learned in the classroom) is 75.6 on a reading test, the dependent variable. The average score on the reading test for Group 2 (which learned using the computer) is 75.7. The amount of variance in both groups is about equal. As you can see, the difference in score is only one tenth of 1 point (75.6 vs. 75.7), but when a t test for the significance between independent means is applied, the results are significant at the .01 level, indicating that computers do work better than classroom teaching. (Chapters 11 and 12 discuss t tests.)

The difference of .01 is indeed statistically significant, but is it meaningful? Does the improvement in test scores (by such a small margin) provide sufficient rationale for the $300,000 it costs to fund the program? Or is the difference negligible enough that it can be ignored, even if it is statistically significant?

Here are some conclusions about the importance of statistical significance that we can reach given this and the countless other possible examples.

- Statistical significance, in and of itself, is not very meaningful unless the study that is conducted has a sound conceptual base that lends some meaning to the significance of the outcome.
- Statistical significance cannot be interpreted independently of the context within which it occurs. For example, if you are the superintendent in a school system, are you willing to retain children in Grade 1 if the retention program significantly raises their standardized test scores by one half point?
- Although statistical significance is important as a concept, it is not the end-all and certainly should not be the only goal of scientific research. That is the reason why we set out to test hypotheses rather than prove them. If our study is designed correctly, then even null results tell you something very important. If a particular treatment does not work, it is important information that others need to know about. If your study is designed well, then you should know why the treatment does not work, and the next person down the line can design his or her study taking into account the valuable information you provided.

Researchers treat the reporting of statistical significance in many different ways in their written reports. Some use words such as significant (assuming that if something is significant it is statistically so), or the entire phrase statistically significant. But some also use the phrase marginally significant, where the probability associated with a finding might be .051 or .053. What to do? You're the boss, if it's your own data being analyzed or if you are reviewing someone else's. Use your noodle and consider all the dimensions of the work being done. If .051, within the context of the question being asked and answered, is "good enough," then it is. Whether outside reviewers agree is a source of great debate and a good topic for class discussion.

AN INTRODUCTION TO INFERENTIAL STATISTICS

Whereas descriptive statistics are used to describe a sample's characteristics, **inferential statistics** are used to infer something about the population based on the sample's characteristics.

At several points throughout the first half of *Statistics for People Who (Think They) Hate Statistics*, we have emphasized that a hallmark of good scientific research is choosing a sample in such a way that it is representative of the population from which it was selected. The process then becomes an inferential one, where you infer from the smaller sample to the larger population based on the results of tests (and experiments) conducted using the sample.

Before we start discussing individual inferential tests, let's go through the logic of how the inferential method works.

How Inference Works

Here are the general steps of a research project to see how the process of inference might work. We'll stay with adolescents' attitudes toward mothers working as an example.

Here's the sequence of events that might happen.

1. The researcher selects representative samples of adolescents who have mothers who work and adolescents who have mothers who do not work. These are selected in such a way that the samples represent the populations from which they are drawn.

2. Each adolescent is administered a test to assess his or her attitude. The mean scores for groups are computed and compared using some test.

3. A conclusion is reached as to whether or not the difference between the scores is the result of chance (meaning some factor other than moms working is responsible for the difference), or the result of "true" and statistically significant differences between the two groups (meaning the results are due to moms working).

4. A conclusion is reached as to the relationship between maternal employment and adolescents' attitudes in the population from which the sample was originally drawn. In other words, an inference, based on the results of an analysis of the sample data, is made about the population of all adolescents.

How to Select What Test to Use

Step 3 above brings us to ask the question, "How do I select the appropriate statistical test to determine if a difference between groups exists?" Heaven knows, there are plenty of them, and you have to decide which one to use and when to use it. Well, the best way to learn which test to use is to be an experienced statistician who has taken lots of courses in this area and participated in lots of research. Experience is still the greatest teacher. In fact, there's no way you can really *learn* what to use and when to use it unless you've had the real-life, applied opportunity to actually use these tools. And as a result of taking this course, you are learning how to use these very tools.

So, for our purposes and to get started, we've created this nice little flow chart (a.k.a. cheat sheet) of sorts that you see in Figure 9.1 on page 174. You have to have some idea of what you're doing, so selecting the correct statistical test is not entirely autopilot, but it certainly is a good place to get started.

Don't think for a second that Figure 9.1 takes the place of your need to learn about when these different tests are appropriate. The flow chart is here only to help you get started.

Here's How to Use the Chart

1. Assume that you're very new to this statistics stuff (which you are) and that you have some idea of what these tests of significance are, but you're pretty lost as far as deciding which one to use when.

2. Answer the question at the top of the flow chart.

3. Proceed down the chart by answering each of the questions until you get to the end of the chart. That's the statistical test you should use. This is not rocket science, and with some practice (which you will get throughout this part of *Statistics for People . . .*), you'll be able to quickly and reliably select the appropriate test. Each of the chapters in this part of the book will begin with a chart like the one you see in Figure 9.1 and take you through the specific steps for the test statistic you should use.

Does the cute flow chart in Figure 9.1 contain all the statistical tests there are? Not by a long shot. There are hundreds, but the ones in Figure 9.1 are the ones used most often. And if you are going to become familiar with the research in your own field, you are bound to run into these.

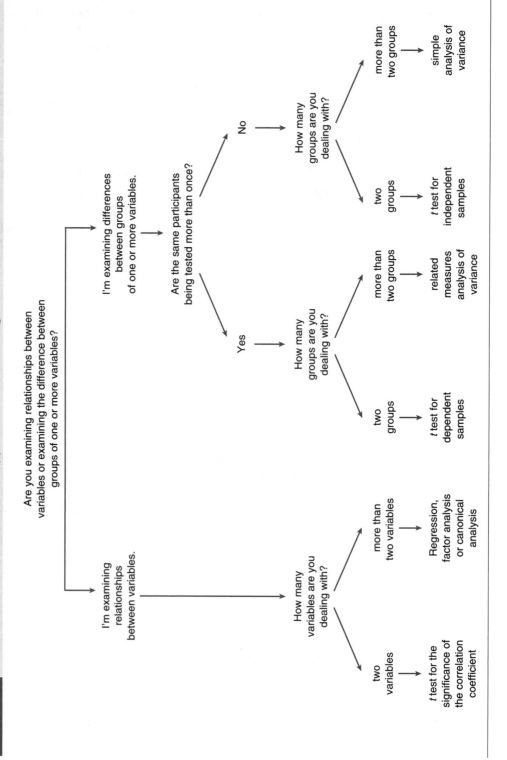

AN INTRODUCTION TO TESTS OF SIGNIFICANCE

What inferential statistics does best is allow decisions to be made about populations based on the information about samples. One of the most useful tools for doing this is a test of statistical significance that can be applied to different types of situations, depending on the nature of the question being asked and the form of the null hypothesis.

For example, do you want to look at the difference between two groups, such as whether boys score significantly differently from girls on some test? Or the relationship between two variables, such as number of children in a family and average score on intelligence tests? The two cases call for different approaches, but both will result in a test of a null hypothesis using a specific test of statistical significance.

How a Test of Significance Works: The Plan

Tests of significance are based on the fact that each type of null hypothesis has associated with it a particular type of statistic. And each of the statistics has associated with it a special distribution that you compare with the data you obtain from a sample. A comparison between the characteristics of your sample and the characteristics of the test distribution allows you to conclude if the sample characteristics are different from what you would expect by chance.

Here are the general steps to take in the application of a statistical test to any null hypothesis. These steps will serve as a model for each of the chapters in Part IV.

1. A statement of the null hypothesis. Do you remember that the null hypothesis is a statement of equality? The null hypothesis is the "true" state of affairs given no other information on which to make a judgment.

2. Setting the level of risk (or the level of significance or Type I error) associated with the null hypothesis. With any research hypothesis comes a certain degree of risk that you are wrong. The smaller this error is (such as .01 compared with .05), the less risk you are willing to take. No test of a hypothesis is completely risk free because you never really know the "true" relationship between variables. Remember that it is traditional to set the Type I error rate at .01 or .05; SPSS and other programs specify the exact level.

3. Selection of the appropriate test statistic. Each null hypothesis has associated with it a particular test statistic. You can learn what test is related to what type of question in this part of *Statistics for People* . . .

4. Computation of the test statistic value. The **test statistic value** (called the **obtained value**) is the result of a specific statistical test. For example, there are test statistics for the significance of the difference between the averages of two groups, for the significance of the difference of a correlation coefficient from 0, and for the significance of the difference between two proportions. You'll actually compute the test statistic and come up with a numerical value.

5. Determination of the value needed for rejection of the null hypothesis using the appropriate table of critical values for the particular statistic. Each test statistic (along with group size and the risk you are willing to take) has a **critical value** associated with it. This is the value you would expect the test statistic to yield if the null hypothesis is indeed true. Now, remember that more and more computer applications that analyze data report the exact probability associated with the test statistic, so once you know your way around this block, this step (and several that follow) becomes much less important.

6. Comparison of the obtained value to the critical value. This is the crucial step. Here, the value you obtained from the test statistic (the one you computed) is compared with the value (the critical value) you would expect to find by chance alone.

7. If the obtained value is more extreme than the critical value, the null hypothesis cannot be accepted. That is, the null hypothesis's statement of equality (reflecting chance) is not the most attractive explanation for differences that were found. Here is where the real beauty of the inferential method shines through. Only if your obtained value is more extreme than chance (meaning that the result of the test statistic is not a result of some chance fluctuation) can you say that any differences you obtained are not due to chance and that the equality stated by the null hypothesis is not the most attractive explanation for any differences you might have found. Instead, the differences must be due to the treatment.

8. If the obtained value does not exceed the critical value, the null hypothesis is the most attractive explanation. If you cannot show that the difference you obtained is due to something other than chance (such as the treatment), then the difference must be

due to chance or something you have no control over. In other words, the null is the best explanation.

Here's the Picture That's Worth a Thousand Words

What you see in Figure 9.2 represents the eight steps that we just went through. This is a visual representation of what happens when the obtained and critical values are compared. In this example, the significance level is set at .05, or 5%. It could have been set at .01, or 1%.

In examining Figure 9.2, note the following:

1. The entire curve represents all the possible outcomes based on a specific null hypothesis, such as the difference between two groups or the significance of a correlation coefficient.

2. The critical value is the point beyond which the obtained outcomes are judged to be so rare that the conclusion is that the obtained outcome is not due to chance but to some other factor. In this example, we define rare as having a less than 5% chance of occurring.

3. If the outcome representing the obtained value falls to the left

Figure 9.2	Comparing Obtained Values to Critical Values and Making Decisions About Rejecting or Accepting the Null Hypothesis

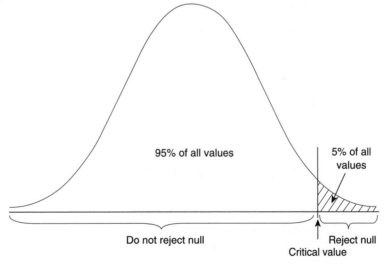

of the critical value (it is less extreme), the conclusion is that the null hypothesis is the most attractive explanation for any differences that are observed. In other words, the obtained value falls in the region (95% of the area under the curve) where we expect only outcomes due to chance to occur.

4. If the obtained value falls to the right of the critical value (it is more extreme), the conclusion is that the research hypothesis is the most attractive explanation for any differences that are observed. In other words, the obtained value falls in the region (5% of the area under the curve) where we would expect only outcomes due to something other than chance to occur.

BE EVEN MORE CONFIDENT

You now know that probabilities can be associated with outcomes—that's been an ongoing theme for the past two chapters. Now we are going to say the same thing in a bit of a different way and also introduce a new idea called confidence intervals.

A **confidence interval** is the best estimate of the range of a population value (or population parameter) that we can come up with given the sample value (or sample statistic). For example, if we know the mean spelling score for a sample of 20 third graders (of all the third graders in a school district), how much confidence can we have that the population mean will fall between two scores? So, for example, a 95% confidence interval would be correct (defined as the sample statistic representing the population parameter) 95% of the time.

You already know that the probability of a raw score falling within ± 1.96 z scores or standard deviations is 95%, right (and see page 153 in Chapter 8 if you need some review)? Or, the probability of a raw score falling within ± 2.56 z scores or standard deviations is 99%. If we use the plus or minus raw scores equivalent to those z scores, we have a confidence interval.

Let's fool around with some real numbers.

Let's say that the mean spelling score for a random sample of 100 sixth graders is 64 (out of 75 words) and the standard deviation is 5. What confidence can we have in predicting the population mean for the average spelling score for the entire population of sixth graders?

The 95% confidence interval is equal to . . .

$$64 \pm 1.96(5)$$

or a range from 54.2 to 73.8, so at the least you can say with 95% confidence that the population mean for the average spelling score for all sixth graders falls between those two scores.

Want to be more confident? The 99% confidence interval would be computed as follows . . .

$$64 \pm 2.56(5)$$

or a range from 51.2 to 76.8, so you can conclude with 99% confidence that the population mean falls between those two scores.

Why does the confidence interval itself get larger as the probability of your being correct increases (from, say, 95% to 99%)? Because the larger range of the confidence interval [in this case from 19.5 (73.8 – 54.2) for a 95% confidence interval to 25.6 (76.8 – 51.2) for a 99% confidence interval] allows you to encompass a larger amount of possible outcomes and you can thereby be more confident. Ha! Isn't this stuff cool?

SUMMARY

So, now you know exactly how the concept of significance works, and all that is left is applying it to a variety of different research questions. That's what we'll start with in the next chapter and continue with through most of this part of the book.

TIME TO PRACTICE

1. Why is significance an important construct in the study and use of inferential statistics?

2. What does the (idea of the) critical value represent?

3. Given the following information, would your decision be to reject or fail to reject the null hypothesis? Setting the level of significance at .05 for decision making, provide an explanation for your conclusion.

   a. The null hypothesis that there is no relationship between the type of music a person listens to and his crime rate ($p < .05$).

   b. The null hypothesis that there is no relationship between the amount of coffee consumption and GPA ($p = .62$).

   c. The null hypothesis that there is a negative relationship between the number of hours worked and level of job satisfaction ($p = .51$).

4. What's wrong with the following statements?

   a. A Type I error of .05 means that 5 times out of 100, I will reject a true null hypothesis.

   b. It is possible to set the Type I error rate to 0.

   c. The smaller the Type I error rate, the better the results.

5. Why is it "harder" to find a significant outcome (all other things being equal) when the research hypothesis is being tested at the .01 rather than the .05 level of significance?

6. Why should we think in terms of "failing to reject" the null, rather than just accepting it?

7. Here's more on the significance-meaningfulness debate.

   a. Provide an example where a finding may be statistically significant and meaningful.

   b. Now provide an example where a finding may be statistically significant and not meaningful.

8. What does chance have to do with testing the research hypothesis for significance?

9. In Figure 9.2 (p. 177), there is a striped area on the right side of the illustration.

   a. What does that entire striped area represent?

   b. If the striped area were a larger portion underneath the curve, what would that represent?

10 Only the Lonely

The One-Sample Z Test

Difficulty Scale ☺☺☺ (not too hard—this is the first one of this kind, but you know more than enough to master it)

WHAT YOU'LL LEARN ABOUT IN THIS CHAPTER

✦ When the Z test for one sample is appropriate to use

✦ How to compute the observed z value

✦ Interpreting the z value and understanding what it means

INTRODUCTION TO THE ONE-SAMPLE Z TEST

Lack of sleep can cause all kinds of problems, from grouchiness to fatigue and, in rare cases, even death. So, you can imagine health care professionals' interest in seeing that their patients get enough sleep. And, this is especially the case for those who are ill and have a real need for the healing and rejuvenating qualities that sleep brings. Dr. Joseph Cappelleri and his colleagues looked at the sleep difficulties of patients with a particular illness, fibromyalgia, to evaluate the usefulness of the Medical Outcomes Study (MOS) Sleep Scale as a measure of sleep problems. Although other analyses were completed, including one that compared a treatment and a control group to one another, the important analysis (for our discussion) was the comparison of participants' MOS scores to national MOS norms. Such a comparison between a sample (the value of the MOS score for participants in this study) to a population's score (the norms) necessitates the use of a one-sample Z test.

And their findings? MOS Sleep Scale scores were statistically ($p < .001$) lower than the average for the population (meaning they did not sleep as well), which means that the sample being tested did not have the same characteristics (at least on this measure of sleep) as did the general population. In other words, the null hypothesis that the sample average and the population average were equal could not be rejected.

So why use the one-sample Z test? Cappelleri and his colleagues wanted to know if the sample values were different from population (national) values collected using the same measure. They were, in effect, comparing a sample statistic to a population parameter and seeing if they could conclude that the sample was (or was not) representative of the population.

Want to know more? Check out Cappelleri, J., Bushmakin, A., McDermott, A., Dukes, E., Sadosky, A., Petrie, C., & Martin, S. (2009). Measurement properties of the Medical Outcomes Study Sleep Scale in patients with fibromyalgia. *Sleep Medicine, 10*(7), 766–770.

The Path to Wisdom and Knowledge

Here's how you can use Figure 10.1, the flow chart introduced in Chapter 9, to select the appropriate test statistic, the **one-sample Z test**. Follow along the highlighted sequence of steps in Figure 10.1. Now this is pretty easy (and they are not all this easy) because this is the only inferential procedure in all of Part IV of *Statistics for People* . . . where we have only one group. And, there's lots of stuff here that will take you back to Chapter 8 and standard scores. And, because you're an expert on those . . .

1. We are examining differences between a sample and a population.

2. There is only one group being tested.

3. The appropriate test statistic is a one-sample Z test.

COMPUTING THE TEST STATISTIC

The formula used for computing the value for the one-sample Z test is shown in Formula 10.1. Remember that we are testing whether a sample mean belongs to or represents a population mean. The difference between the sample mean ($\bar{X}$) and the population (μ) mean

Figure 10.1 Determining That a One-Sample Z Test Is the Correct Statistic

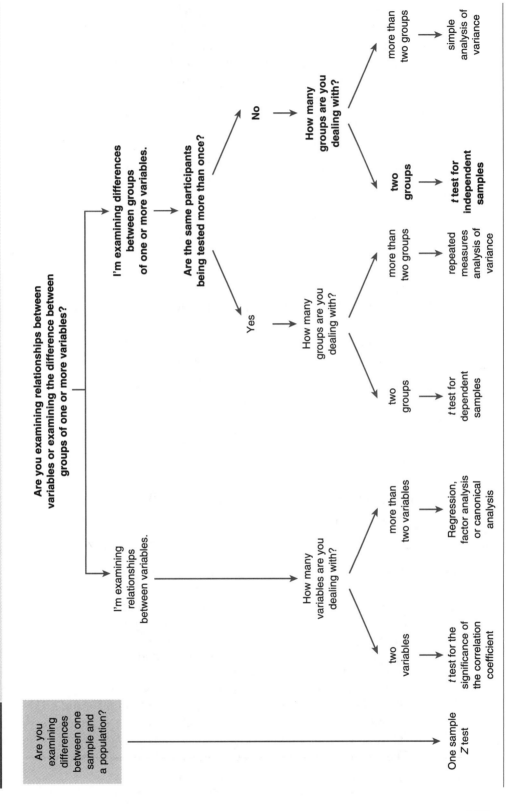

makes up the numerator for the z value. The denominator is called the standard error of the mean, an error term, and is the value we would expect by chance given all the variability that surrounds the selection of all possible sample means from a population. Using this standard error of the mean (and the key term here is *standard*) allows us once again (as we showed in Chapter 9) to use the table of z scores to reach a decision as to the probability of an outcome.

$$z = \frac{\overline{X} - \mu}{SEM}$$ (10.1)

where

$\overline{X}$ = the mean of the sample

μ = the population average

SEM = the standard error of the mean

Now, to compute the standard error of the mean, which you need in Formula 10.1, use the following formula . . .

$$SEM = \frac{\sigma}{\sqrt{n}}$$

where

σ = the standard deviation for the population

n = the size of the sample

Tech Talk

The standard error of the mean is the standard deviation of all the possible means selected from the population. It's the best *estimate* we can come up with given that it is impossible to compute *all* the possible means. If our sample selection were perfect, the difference between the sample and the population averages would be zero, right? Right. If the sampling from a population were not done correctly (randomly and representatively), then the standard deviation of all these samples could be huge, right? Right. But, even though we try, we just can't seem to select the perfect sample; there's always some error, and the standard error of the mean reflects what that value would be for the entire population of all mean values. And yes, Virginia, this is the standard error of the mean. There can be (and are) standard errors for other measures as well.

Time for an example.

Dr. McDonald thinks that his group of earth science students is particularly special (in a good way), and he is interested in knowing if their class average falls within the boundaries of the average score for the larger group of students who have taken earth science over the past 20 years. Because he's kept good records, he knows the means and standard deviations for both his group of 36 students and the larger group of 1,000 past enrollees. Here are the data he has, and the values for which we will need to compute the z score are italicized.

	Size	**Mean**	**Standard Deviation**
Sample	36	100	5
Population	1,000	99	2.5

Here are the famous eight steps and the computation of the Z-test statistic.

1. A statement of the null and research hypotheses.

The null hypothesis states that sample average is equal to the population average. If the null is not rejected, it means that the sample is representative of the population. If the null is rejected in favor of the research hypothesis, it means that the sample average is different from the population average.

The null hypothesis is . . .

$$H_0 : \overline{X} = \mu \qquad (10.2)$$

The research hypothesis in this example is . . .

$$H_1 : \overline{X} \neq \mu \qquad (10.3)$$

2. Setting the level of risk (or the level of significance or Type I error) associated with the null hypothesis.

The level of risk or Type I error or level of significance (any other names?) is .05, totally the decision of the researcher.

3. Selection of the appropriate test statistic.

Using the flow chart shown in Figure 10.1, we determined that the appropriate test is a one-sample Z test.

    4. Computation of the test statistic value (called the obtained value).

Now's your chance to plug in values and do some computation. The formula for the z value was shown in Formula 10.1. The specific values are plugged in (first for the SEM in Formula 10.4 and then for z in Formula 10.5). (We got all of these data from the table you saw on page 185.) With the values plugged in, we get the following results.

$$SEM = \frac{2.5}{\sqrt{36}} = .42 \qquad\qquad (10.4)$$

$$z = \frac{100 - 99}{.42} = 2.38 \qquad\qquad (10.5)$$

The z value for a comparison of the sample mean to this population mean given the data on page 185 is 2.38.

    5. Determination of the value needed for rejection of the null hypothesis using the appropriate table of critical values for the particular statistic.

Here's where we go to Table B.1 in Appendix B, which lists the probabilities associated with specific z values, which are the critical values for the rejection of the null hypothesis. This is just as we did in Chapter 9 with several examples.

We can use the values in Table B.1 to see if two means "belong" to one another by comparing what we would expect by chance (the tabled or critical value) to what we observe (the obtained value).

From our work in Chapter 9, we know that a z value of +1.96 has associated with it a probability of .025, and if we consider that the sample mean could be bigger, or smaller, than the population mean, we need to consider both ends of the distribution (and a range of ±1.96) and a total Type I error rate of .05.

    6. A comparison of the obtained value and the critical value.

The obtained z value is 2.38. So, for a test of this null hypothesis at the .05 level with 36 participants, the critical value is ±1.96. This value represents the value at which chance is the most attractive explanation as to why the sample mean and the population mean differ. Beyond that critical value in either direction (remember that the research hypothesis is nondirectional and this is a two-tailed test) means that we need to provide an explanation as to why the sample and the population means differ.

7. and 8. Decision time!

If the obtained value is more extreme than the critical value (remember Figure 9.2), the null hypothesis cannot be accepted. If the obtained value does not exceed the critical value, the null hypothesis is the most attractive explanation. In this case, the obtained value (2.38) does exceed the critical value (1.96), and it is absolutely extreme enough for us to say that the sample of 36 students in Dr. Kent's class is different from the previous 1,000 students who have also taken the course. If the obtained value were less than 1.96, it would mean that there is no difference between the test performance of the sample and the 1,000 students who have taken the test over the past 20 years. In this case, the 36 students performed basically at the same level as the previous 1,000.

And, the final step? Why, of course. Why does this group of students differ? Perhaps Kent is right in that they are smarter, but they may also be better users of technology or more motivated. Perhaps they just studied harder. All questions to be tested some other time.

So How Do I Interpret $z = 2.38$, $p < .05$?

- z represents the test statistic that was used.
- 2.38 is the obtained value, obtained using the formulas we showed you earlier in the chapter.
- $p < .05$ (the really important part of this little phrase) indicates that the probability is less than 5% that on any one test of the null hypothesis, the sample and the population averages differ.

SUMMARY

The one-sample Z test is the most simple example of an inferential test, and that's why we started off this long section with an explanation of what this test does and how it is applied. That's why we started this new edition with this test—a simple and direct introduction. But, the (very) good news is that most (if not all) of the steps we take as we move to more complex analytic tools are exactly the same as what you saw here. Let's move on to a very common inferential test (and an extension of the Z test we covered here), the simple *t* test between the means of two different groups.

TIME TO PRACTICE

1. When is it appropriate to use the one-sample Z test?

2. What's with the z in Z test? What similarity does it have to a simple z or standard score?

3. For the following situations, write out in words a research hypothesis.
   a. Bob wants to know if the weight loss for his group on the chocolate-only diet is representative of weight loss in a large population of middle-aged men.
   b. The health department is charged with finding out if the rate of flu per thousand citizens for this past flu season is comparable to the average rate of the past 50 seasons.
   c. Blair is almost sure that his monthly costs for the past year are not representative of his average monthly costs over the past 20 years.

4. Flu cases this past flu season in the Oshkosh school system were about 15 per week. For the entire state, the weekly average is 16 and the standard deviation, 2.35. Are the kids in Oshkosh as sick as the kids throughout the state?

11

t(ea) for Two

Tests Between the Means of Different Groups

Difficulty Scale ☺☺☺ A little longer than the previous chapter, but basically the same kind of procedures and very similar questions. Not too hard, but you have to pay attention.

WHAT YOU'LL LEARN ABOUT IN THIS CHAPTER

✦ When the t test for independent means is appropriate to use

✦ How to compute the observed t value

✦ Interpreting the t value and understanding what it means

INTRODUCTION TO THE T TEST
FOR INDEPENDENT SAMPLES

Even though eating disorders are recognized for their seriousness, little research has been done that compares the prevalence and intensity of symptoms across different cultures. John P. Sjostedt, John F. Shumaker, and S. S. Nathawat undertook this comparison with groups of 297 Australian and 249 Indian university students. Each student was tested on the Eating Attitudes Test and the Goldfarb Fear of Fat Scale. The groups' scores were compared with one another. On a comparison of means between the Indian and the Australian participants, Indian students scored higher on both of the tests. The results for the Eating Attitudes Test were $t_{(524)} = -4.19$,

$p < .0001$, and the results for the Goldfarb Fear of Fat Scale were $t_{(524)} = -7.64$, $p < .0001$.

Now just what does all this mean? Read on.

Why was the t test for independent means used? Sjostedt and his colleagues were interested in finding out if there was a difference in the average scores of one (or more) variable(s) between the two groups that were independent of one another. By *independent,* we mean that the two groups were not related in any way. Each participant in the study was tested only once. The researchers applied a t test for independent means, arriving at the conclusion that for each of the outcome variables, the differences between the two groups were significant at or beyond the .0001 level. Such a small Type I error means that there is very little chance that the difference in scores between the two groups was due to something other than group membership, in this case representing nationality, culture, or ethnicity.

Want to know more? Check out Sjostedt, J. P., Shumaker, J. F., & Nathawat, S. S. (1998). Eating disorders among Indian and Australian university students. *Journal of Social Psychology, 138*(3), 351–357.

The Path to Wisdom and Knowledge

Here's how you can use Figure 11.1, the flow chart introduced in Chapter 9, to select the appropriate test statistic, the t test for independent means. Follow along the highlighted sequence of steps in Figure 11.1.

1. The differences between the groups of Australian and Indian students are being explored.

2. Participants are being tested only once.

3. There are two groups.

4. The appropriate test statistic is t test for independent means.

Almost every statistical test has certain assumptions that underlie the use of the test. For example, the t test has a major assumption that the amount of variability in each of the two groups is equal. This is the homogeneity of variance assumption. Although this assumption can be violated if the sample size is big enough, small samples and a violation

Figure 11.1

Determining That a *t* Test Is the Correct Statistic Here

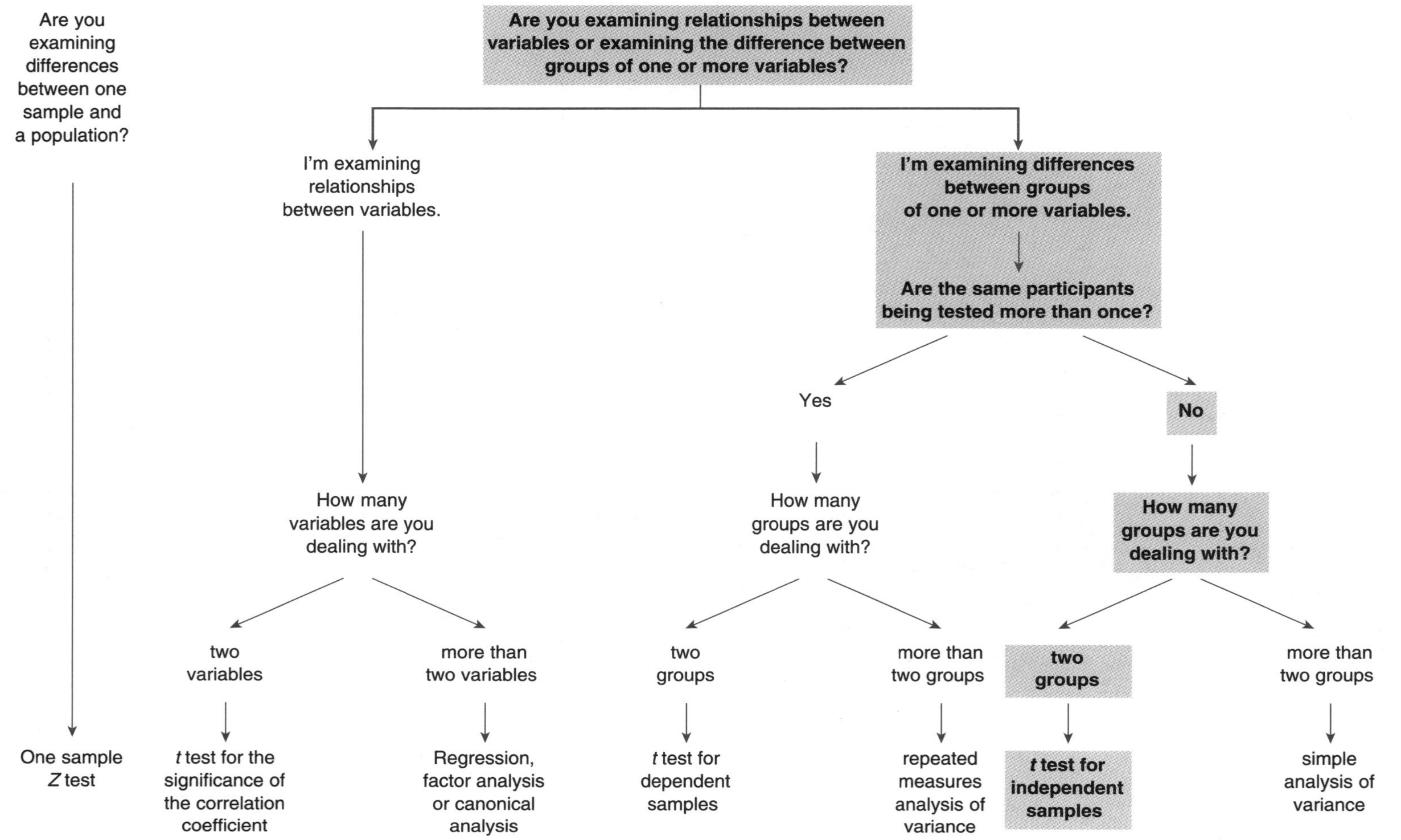

of this assumption can lead to ambiguous results and conclusions. Don't knock yourself out worrying about these assumptions because they are beyond the scope of this book. However, you should know that such assumptions are rarely violated, but it is worth knowing that they do exist.

> As we mentioned earlier, there are hundreds of statistical tests, and the only inferential one that used one sample that we cover in this book is the one sample Z test (see Chapter 10). But, there is also the one-sample t test that compares the mean score of a sample to another score, and sometimes that score is, indeed, the population mean, just as with the one-sample Z test. In any case, you can use the one-sample z or one-sample t test to test the same hypothesis and you will reach the same conclusions (although you will be using different values and tables to do so).

COMPUTING THE TEST STATISTIC

The formula for computing the t value for the t test for independent means is shown in Formula 11.1. The difference between the means makes up the numerator of the following formula used to compute the t value or the test statistic of the obtained value. The amount of variation within and between each of the two groups makes up the denominator.

$$t = \frac{\overline{X}_1 - \overline{X}_2}{\sqrt{\left[\frac{(n_1-1)s_1^2 + (n_1-1)s_2^2}{n_1+n_2-2}\right]\left[\frac{n_1+n_2}{n_1 n_2}\right]}}$$ (11.1)

where

$\overline{X}_1$ is the mean for Group 1

$\overline{X}_2$ is the mean for Group 2

n_1 is the number of participants in Group 1

n_2 is the number of participants in Group 2

s_1^2 is the variance for Group 1

s_2^2 is the variance for Group 2

Nothing new here at all. It's just a matter of plugging in the correct values.

Here are some data reflecting the number of words remembered following a program designed to help Alzheimer's patients remember the order of daily tasks. Group 1 was taught using visuals, and Group 2 was taught using visuals and intense verbal rehearsal. We'll use the data to compute the test statistic in the following example.

Group 1			Group 2		
7	5	5	5	3	4
3	4	7	4	2	3
3	6	1	4	5	2
2	10	9	5	4	7
3	10	2	5	4	6
8	5	5	7	6	2
8	1	2	8	7	8
5	1	12	8	7	9
8	4	15	9	5	7
5	3	4	8	6	6

Here are the famous eight steps and the computation of the *t*-test statistic.

1. A statement of the null and research hypotheses.

As represented by Formula 11.2, the null hypothesis states that there is no difference between the means for Group 1 and Group 2. For our purposes, the research hypothesis (shown as Formula 11.3) states that there is a difference between the means of the two groups. The research hypothesis is a two-tailed, nondirectional research hypothesis because it posits a difference, but in no particular direction.

The null hypothesis is

$$H_0: \mu_1 = \mu_2 \qquad\qquad (11.2)$$

The research hypothesis is

$$H_1 : \overline{X}_1 \neq \overline{X}_2 \qquad (11.3)$$

2. Setting the level of risk (or the level of significance or Type I error) associated with the null hypothesis.

The level of risk or Type I error or level of significance (any other names?) is .05, totally the decision of the researcher.

3. Selection of the appropriate test statistic.

Using the flow chart shown in Figure 11.1, we determined that the appropriate test is a t test for independent means. It is not a t test for dependent means (a common mistake beginning students make) because the groups are independent of one another.

4. Computation of the test statistic value (called the obtained value).

Now's your chance to plug in values and do some computation. The formula for the t value was shown in Formula 11.1. When the specific values are plugged in, we get the equation shown in Formula 11.4. (We already computed the mean and standard deviation.)

$$t = \frac{5.43 - 5.53}{\sqrt{\left[\frac{(30-1)3.42^2 + (30-1)2.06^2}{30+30-2}\right]\left[\frac{30+30}{30\times30}\right]}} \qquad (11.4)$$

With the numbers plugged in, Formula 11.5 shows how we got the final value of −.137. The value is negative because a larger value (the mean of Group 2, which is 5.53) is being subtracted from a smaller number (the mean of Group 1, which is 5.43). Remember, though, that because the test is nondirectional and any difference is hypothesized, the sign of the difference is meaningless.

$$t = \frac{-.1}{\sqrt{\left(\frac{339.20 + 123.06}{58}\right)\left(\frac{60}{900}\right)}} = -.137 \qquad (11.5)$$

5. Determination of the value needed for rejection of the null hypothesis using the appropriate table of critical values for the particular statistic.

Here's where we go to Table B.2 in Appendix B, which lists the critical values for the *t* test.

We can use this distribution to see if two independent means differ from one another by comparing what we would expect by chance (the tabled or critical value) to what we observe (the obtained value).

Our first task is to determine the **degrees of freedom** (*df*), which approximate the sample size. For this particular test statistic, the degrees of freedom are $n_1 - 1 + n_2 - 1$. So for each group, add the size of the two samples and subtract 2. In this example, $30 + 30 - 2 = 58$. These are the degrees of freedom for this test statistic and not necessarily for any other.

Using this number (58), the level of risk you are willing to take (earlier defined as .05), and a two-tailed test (because there is no direction to the research hypothesis), you can use the *t*-test table to look up the critical value. At the .05 level, with 58 degrees of freedom for a two-tailed test, the value needed for rejection of the null hypothesis is . . . Oops! There's no 58 degrees of freedom in the table! What do you do? Well, if you select the value that corresponds to 55, you're being conservative in that you are using a value for a sample smaller than what you have (and the critical *t* value will be larger).

If you go for 60 degrees of freedom (the closest to your value of 58), you will be closer to the size of the population, but a bit liberal in that 60 is larger than 58. Although statisticians differ in their viewpoint as to what to do in this situation, let's always go with the value that's closest to the actual sample size. So the value needed to reject the null hypothesis with 58 degrees of freedom at the .05 level of significance is 2.001.

6. A comparison of the obtained value and the critical value.

The obtained value is −.14, and the critical value for rejection of the null hypothesis that Group 1 and Group 2 performed differently is 2.001. The critical value of 2.001 represents the value at which chance is the most attractive explanation for any of the observed differences between the two groups given 30 participants in each group and the willingness to take a .05 level of risk.

7. and 8. Decision time!

Now comes our decision. If the obtained value is more extreme than the critical value (remember Figure 9.2), the null hypothesis cannot be accepted. If the obtained value does not exceed the critical value, the null hypothesis is the most attractive explanation. In this case, the obtained value (–.14) does not exceed the critical value (2.001)—it is not extreme enough for us to say that the difference between Groups 1 and 2 occurred by anything other than chance. If the value were greater than 2.001, it would represent a value that is just like getting 8, 9, or 10 heads in a coin toss—too extreme for us to believe that something else other than chance is not going on. In the case of the coin, it's an unfair coin—in this example, it would be that there is a better way to teach memory skills to these older people.

So, to what can we attribute the small difference between the two groups? If we stick with our current argument, then we could say the difference is due to anything from sampling error to rounding error to simple variability in participants' scores. Most important, we're pretty sure (but, of course, not 100% sure—that's what level of significance and Type I errors are all about, right?) that the difference is not due to anything in particular that one group or the other experienced to make its scores better.

So How Do I Interpret $t_{(58)} = -.14, p > .05$?

- t represents the test statistic that was used
- 58 is the number of degrees of freedom
- –.14 is the obtained value, obtained using the formula we showed you earlier in the chapter
- $p > .05$ (the really important part of this little phrase) indicates that the probability is greater than 5% that on any one test of the null hypothesis, the two groups do not differ because of the way they were taught. Also note that the $p > .05$ can also appear as p = n.s. for nonsignificant.

SPECIAL EFFECTS: ARE THOSE DIFFERENCES FOR REAL?

OK, now you have some idea how to test for the difference between the averages of two separate or independent groups. Good job. But that's not the whole story.

You may have a significant difference between groups, but the $64,000 question is not only whether that difference is (statistically) significant, but also whether it is *meaningful*. We mean, is there enough of a separation between the distribution that represents each group so that the difference you observe and the difference you test is really a difference! Hmm. . . . Welcome to the world of effect size.

Effect size is a measure of how different two groups are from one another—it's a measure of the magnitude of the treatment. Kind of like how big is big. And what's especially interesting about computing effect size is that sample size is not taken into account. Calculating effect size, and making a judgment about it, adds a whole new dimension to understanding significant outcomes.

Let's take the following example. A researcher tests the question of whether participation in community-sponsored services (such as card games, field trips, etc.) increases the quality of life (as rated from 1 to 10) for older Americans. The researcher implements the treatment over a 6-month period and then, at the end of the treatment period, measures quality of life in the two groups (each consisting of 50 participants over the age of 80 where one group got the services and one group did not). Here are the results.

	No Community Services	**Community Services**
Mean	6.90	7.46
Standard Deviation	1.03	1.53

And, the verdict is that the difference is significant at the .034 level (which is $p < .05$, right?).

OK, there's a significant difference, but what about the magnitude of the difference?

The great Pooh-bah of effect size was Jacob Cohen, who wrote some of the most influential and important articles on this topic. He authored a very important and influential book (your stat teacher has it on his or her shelf!) that instructs researchers how to figure out the effect size for a variety of different questions that are asked about differences and relationships between variables. Here's how you do it.

Computing and Understanding the Effect Size

Just as with many other statistical techniques, there are many different ways to compute the effect size. We are going to show you the most simple and straightforward. You can learn more about effect sizes in some of the references we'll be giving you in a minute.

By far, the most direct and simple way to compute effect size is to simply divide the difference between the means by any one of the standard deviations. Danger, Will Robinson—this does assume that the standard deviations (and the amount of variance) between groups are equal to one another.

For our example above, we'll do this . . .

$$ES = \frac{\overline{X}_1 - \overline{X}_2}{SD}$$

where

ES = effect size

$\overline{X}_1$ = the mean for Group 1

$\overline{X}_2$ = the mean for Group 2

SD = the standard deviation from either group

So, in our example,

$$ES = \frac{7.46 - 6.90}{1.53}$$

or .366. So, the effect size for this example is .37.

What does it mean? One of the very cool things that Cohen (and others) figured out was just what a small, medium, and large effect size is. They used the following guidelines:

A small effect size ranges from 0.0 to .20

A medium effect size ranges from .20 to .50

A large effect size is any value above .50

Our example, with an effect size of .37, is categorized as medium. But what does it really mean?

Effect size gives us an idea about the relative positions of one group to another. For example, if the effect size is 0, that means that both groups tend to be very similar and overlap entirely—there is no difference between the two distributions of scores. On the other hand, an effect size of 1 means that the two groups overlap about 45% (having that much in common). And, as you might expect, as the effect size gets larger, it reflects the increasing lack of overlap between the two groups.

Jacob Cohen's book, *Statistical Power Analysis for the Behavioral Sciences*, first published in 1967 with the latest edition (1988) available from Lawrence Erlbaum Associates (now part of Taylor and Francis), is a must for anyone who wants to go beyond the very general information that is presented here. It is full of tables and techniques for allowing you to understand how a statistically significant finding is only half the story—the other half is the magnitude of that effect.

So, you really want to be cool about this effect size thing. You can do it the simple way, as we just showed you (by subtracting means from one another and dividing by either standard deviation), or you can really wow that good-looking classmate who sits next to you. The grown-up formula for the effect size uses the pooled variance in the numerator of the ES equation that you saw previously. The pooled standard deviation is sort of an average of the standard deviation from Group 1 and the standard deviation from Group 2. Here's the formula:

$$ES = \frac{\overline{X}_1 - \overline{X}_2}{\sqrt{\frac{\sigma_1^2 + \sigma_2^2}{2}}}$$

where

ES = effect size

$\overline{X}_1$ = the mean of Group 1

$\overline{X}_2$ = the mean of Group 2

σ_1^2 = the variance of Group 1

σ_2^2 = the variance of Group 2

If we applied this formula to the same numbers we showed you previously, you'd get a whopping effect size of .43—not very different from .37, which we got using the more direct method shown earlier (and still in the same category of medium size). But this is a more precise method, and one that is well worth knowing about.

A Very Cool Effect Size Calculator

Why not take the A train and just go right to http://www .uccs.edu/~faculty/lbecker/, where statistician Lee Becker from the University of California developed an effect size calculator? With this calculator, you just plug in the values, click Compute, and the program does the rest, as you see in Figure 11.2. Thanks, Dr. Becker!

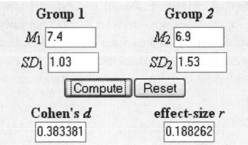

Figure 11.2   The Very Cool Effect Size Calculator

USING THE COMPUTER TO PERFORM A T TEST

SPSS is willing and ready to help you perform these inferential tests. Here's how to perform the one that we just did and interpret the output. We are using the data set named Chapter 11 Data Set 1. From your examination of the data, you can see how the grouping variable (Group 1 or Group 2) is in column 1 and the test variable (memory) is in column 2.

1. Enter the data in the Data Editor or download the file. Be sure that there is a column for group and that you have no more than two groups represented in that column.

2. Click Analyze → Compare Means → Independent-Samples T Test and you will see the Independent-Samples T Test dialog box shown in Figure 11.3.

Figure 11.3   The Dialog Box for Beginning the *t*-Test Analysis

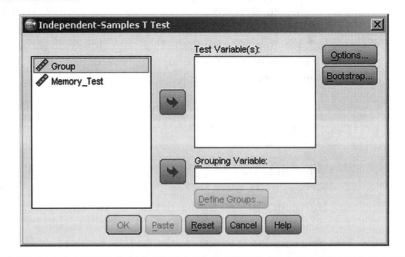

> Notice how SPSS uses a capital *T* to represent this test while we have been using a small *t*? This difference is strictly a matter of personal preference and, more often than not, reflects what people were taught way back when. What's important for you to know is that there is a difference in letter only—it's the same exact test.

3. Click on the variable named Group, and click ➡ to move it to the Grouping Variable(s) box.

4. Click on the variable named Memory_Test, and click ➡ to place it in the Test Variable(s) box.

5. SPSS will not allow you to continue until you define the grouping variable. This basically means telling SPSS how many levels of the group variable there are (wouldn't you think that a program this smart could figure that out?). In any case, click Group (? ?), click Define Groups, and enter the values 1 for Group 1 and 2 for Group 2, as shown in Figure 11.4. The name of the grouping variable (in this case, Group) has to be highlighted before you can define it.

Figure 11.4   The Define Groups Dialog Box

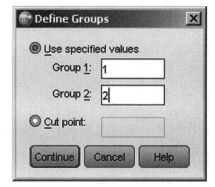

6. Click Continue, and then click OK, and SPSS will conduct the analysis and produce the output you see in Figure 11.5.

Figure 11.5 Copy of SPSS Output for a *t* Test Between Independent Means

T-Test

Group Statistics

	Group	N	Mean	Std. Deviation	Std. Error Mean
Memory_Test	1	30	5.53	3.421	.625
	2	30	5.53	2.063	.377

Independent Samples Test

		Levene's Test for Equality of Variances		*t*-test for Equality of Means					95% Confidence Interval of the Difference	
		F	Sig.	t	df	Sig. (2-tailed)	Mean Difference	Std. Error Difference	Lower	Upper
Memory_Test	Equal variances assumed	4.994	.029	-.137	58	.891	-.100	.729	-1.560	1.360
	Equal variances not assumed			-.137	47.635	.892	-.100	.729	-1.567	1.367

What the SPSS Output Means

There's a ton of SPSS output from this analysis, and for our purposes, we'll deal only with selected output shown in Figure 11.5. There are three things to note.

1. The obtained *t* value is −.137, exactly what we got when we computed the value by hand earlier in this chapter (−.14, rounded up from .1368).

2. The number of degrees of freedom is 58 (which you already know is computed using the formula $n_1 + n_2 - 2$).

3. Here's the really important result. The significance of this finding is .891, or $p = .891$, which means that on one test of this null hypothesis, the likelihood of rejecting the hypothesis when it is true is pretty high (89 out of 100)! So the Type I error is certainly greater than .05, which allowed us to conclude earlier when we did the same analysis using the formula that $p > .05$.

SUMMARY

The *t* test is your first introduction to performing a real statistical test and trying to understand this whole matter of significance from an applied point of view. Be sure that you understand what was in this chapter before you move on. And be sure you can do by hand the few things that were asked for. Next, we move on to using another form of the same test, only this time, there are two measures taken from one group of participants rather than one measure taken from two separate groups.

TIME TO PRACTICE

1. Using the data in the file named Chapter 11 Data Set 2, test the research hypothesis at the .05 level of significance that boys raise their hands in class more often than girls. Do this practice problem by hand using a calculator. What is your conclusion regarding the research hypothesis? Remember to first decide whether this is a one- or two-tailed test.

2. Using the same data set (Chapter 11 Data Set 2), test the research hypothesis at the .01 level of significance that there is a difference between boys and girls in the number of times they raise their hands in class. Do this practice problem by hand using a calculator. What is your conclusion regarding the research hypothesis? You used the same data for this problem as for Question 1, but you have a different

hypothesis (one is directional and the other is nondirectional). How do the results differ and why?

3. Time for some tedious, by-hand practice just to see if you can get the numbers right. Using the following information, calculate the t-test statistic by hand.

a. $\bar{X}_1 = 62$ $\bar{X}_2 = 60$ $n_1 = 10$ $n_2 = 10$ $s_1^2 = 6$ $s_2^2 = 10$

b. $\bar{X}_1 = 158$ $\bar{X}_2 = 157.4$ $n_1 = 22$ $n_2 = 26$ $s_1^2 = 4.23$ $s_2^2 = 6.73$

c. $\bar{X}_1 = 200$ $\bar{X}_2 = 198$ $n_1 = 17$ $n_2 = 17$ $s_1^2 = 6$ $s_2^2 = 5.5$

4. Using the results you got from Question 3 above, and a level of significance at .05, what are the two-tailed critical values associated with each? Would the null hypothesis be rejected?

5. Use the following data and SPSS or some other computer application and write a brief paragraph about whether the in-home counseling is equally effective as the out-of-home treatment for two separate groups. Here are the data. The outcome variable is level of anxiety after treatment on a scale from 1 to 10.

In-Home Treatment	Out-of-Home Treatment
3	7
4	6
1	7
1	8
1	7
3	6
3	5
6	6
5	4
1	2
4	5
5	4
4	3
4	6
3	7
6	5
7	4
7	3
7	8
8	7

6. Using the data in the file named Chapter 11 Data Set 3, test the null hypothesis that urban and rural residents both have the same attitude toward gun control. Use SPSS to complete the analysis for this problem.

7. Here's a good one to think about. A public health researcher tested the hypothesis that providing new car buyers with child safety seats will also act as an incentive for parents to take other measures to protect their children (such as driving more safely, child-proofing the home, etc.). Dr. L counted all the occurrences of safe behaviors in the cars and homes of the parents who accepted the seats versus those who did not. The findings? A significant difference at the .013 level. Another researcher did exactly the same study, and for our purposes, let's assume that everything was the same—same type of sample, same outcome measures, same car seats, and so on. Dr. R's results were marginally significant (remember that from Chapter 9?) at the .051 level. Whose results do you trust more and why?

8. Here are the results of three experiments where the means of the two groups being compared are exactly the same but the standard deviation is quite different from experiment to experiment. Compute the effect size using the formula on page 198 and then discuss why this size changes as a function of the change in variability.

Experiment 1	Group 1 Mean	78.6	Effect Size = _____
	Group 2 Mean	73.4	
	Standard Deviation	2	
Experiment 2	Group 1 Mean	78.6	Effect Size = _____
	Group 2 Mean	73.4	
	Standard Deviation	4	
Experiment 3	Group 1 Mean	78.6	Effect Size = _____
	Group 2 Mean	73.4	
	Standard Deviation	8	

9. Using the data in Chapter 11, Data Set 4 and SPSS, test the null hypothesis that there is no difference in group means between the number of words spelled correctly for two groups of fourth graders. What is your conclusion?

12

t(ea) for Two (Again)

Tests Between the Means of Related Groups

Difficulty Scale ☺☺☺ (not too hard—this is the first one of this kind, but you know more than enough to master it)

INTRODUCTION TO THE T TEST FOR DEPENDENT SAMPLES

How best to educate children is clearly one of the most vexing questions that faces any society. Because children are so different from one another, a balance needs to be found between meeting the basic needs of all while ensuring that special children (on either end of the continuum) get the opportunities they need. An obvious and important part of education is reading, and three professors at the University of Alabama studied the effects of resource and regular classrooms on the reading achievement of learning-disabled children. Renitta Goldman, Gary L. Sapp, and Ann Shumate Foster found that, in general, 1 year of daily instruction in both settings resulted in no difference in overall reading achievement scores. On one specific comparison between the pretest and the posttest of the resource group, they found that $t_{(34)} = 1.23$, $p > .05$. At the beginning of the program, reading achievement scores for children in the resource room were 85.8. At the end of the program, reading achievement scores for children in the resource room were 88.5—a difference, but not a significant one.

So why a test of dependent means? A t test for dependent means indicates that a single group of the same subjects is being studied under two conditions. In this example, the conditions are before the start of the experiment and after its conclusion. Primarily, it is because the same children were tested at two times, before the start of the 1-year program and at the end of the 1-year program, that we use the t test for dependent means. As you can see by the aforementioned result, there was no difference in scores at the beginning and the end of the program. The very small t value (1.23) is not nearly extreme enough to fall outside the region where we would reject the null hypothesis. In other words, there is far too little change for us to say that this difference occurred by something other than chance. The small difference of 2.7 (88.5 − 85.8) is probably due to sampling error or variability within the groups.

Want to know more? Check out Goldman, R., Sapp, G. L., & Foster, A. S. (1998). Reading achievement by learning disabled students in resource and regular classes. *Perceptual and Motor Skills, 86*, 192–194.

The Path to Wisdom and Knowledge

Here's how you can use the flow chart to select the appropriate test statistic, the t test for dependent means. Follow along the highlighted sequence of steps in Figure 12.1.

1. The difference between the students' scores on the pretest and on the posttest is the focus.

2. Participants are being tested more than once.

3. There are two groups of scores.

4. The appropriate test statistic is t test for dependent means.

Figure 12.1 Determining That a *t* Test for Dependent Means Is the Correct Test Statistic

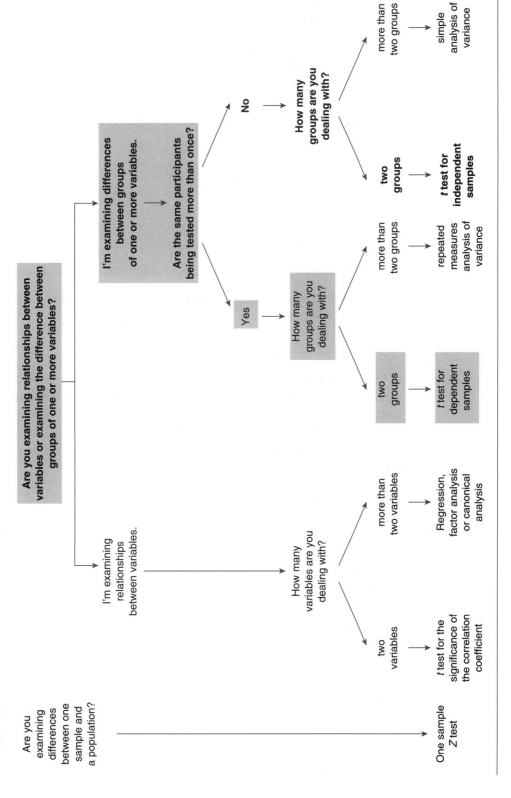

COMPUTING THE TEST STATISTIC

The t test for dependent means involves a comparison of means from each group of scores and focuses on the differences between the scores. As you can see in Formula 12.1, the sum of the differences between the two tests forms the numerator and reflects the difference between groups of scores.

$$t = \frac{\Sigma D}{\sqrt{\frac{n\Sigma D^2 - (\Sigma D)^2}{n-1}}} \qquad (12.1)$$

where

ΣD　is the sum of all the differences between groups of scores

ΣD^2　is the sum of the differences squared between groups of scores

n　is the number of pairs of observations

On the next page some data to illustrate how the t value is computed. Just like in the previous example, there is a pretest and a posttest, and for illustration's sake, assume that these are before and after scores from a reading program.

Here are the famous eight steps and the computation of the t-test statistic.

1. A statement of the null and research hypotheses.

The null hypothesis states that there is no difference between the means for the pretest and the posttest scores on reading achievement. The research hypothesis is a one-tailed, directional research hypothesis because it posits that the posttest score will be higher than the pretest score.

The null hypothesis is

$$H_0: \mu_{\text{posttest}} = \mu_{\text{pretest}} \qquad (12.2)$$

The research hypothesis is

$$H_1: \overline{X}_{\text{posttest}} > \overline{X}_{\text{pretest}} \qquad (12.3)$$

Pretest	Posttest	Difference	D^2
3	7	4	16
5	8	3	9
4	6	2	4
6	7	1	1
5	8	3	9
5	9	4	16
4	6	2	4
5	6	1	1
3	7	4	16
6	8	2	4
7	8	1	1
8	7	−1	1
7	9	2	4
6	10	4	16
7	9	2	4
8	9	1	1
8	8	0	0
9	8	−1	1
9	4	−5	25
8	4	−4	16
7	5	−2	4
7	6	−1	1
6	9	3	9
7	8	1	1
8	12	4	16
Sum	158	188	30	180
Mean	6.32	7.52	1.2	7.2

2. Setting the level of risk (or the level of significance or Type I error) associated with the null hypothesis.

The level of risk or Type I error or level of significance is .05, totally the decision of the researcher.

3. Selection of the appropriate test statistic.

Using the flow chart shown in Figure 12.1, we determined that the appropriate test is a t test for dependent means. It is not a t test for independent means because the groups are not independent of each other. In fact, they're not groups of participants, but groups of scores for the same participants. The groups are dependent on one another. Another name for the t test for dependent means is the t test for paired samples or the t test for correlated samples. You'll see in Chapter 15 that there is a very close relationship between a test of the significance of the correlation between these two sets of scores (pre and post) and the t value we are computing here.

4. Computation of the test statistic value (called the obtained value).

Now's your chance to plug in values and do some computation. The formula for the t value was shown previously. When the specific values are plugged in, we get the equation shown in Formula 12.4. (We already computed the means and standard deviations for the pretest and posttest scores.)

$$t = \frac{30}{\sqrt{\frac{(25 \times 180) - 30^2}{25 - 1}}} \tag{12.4}$$

With the numbers plugged in, we have the following equation with a final obtained t value of 2.45. The mean score for pretest performance was 6.32, and the mean score for posttest performance was 7.52.

$$t = \frac{30}{\sqrt{150}} = 2.45 \tag{12.5}$$

5. Determination of the value needed for rejection of the null hypothesis using the appropriate table of critical values for the particular statistic.

Here's where we go to Table B.2, which lists the critical values for the *t* test. Once again, we have a *t* test and we'll use the same table we used in Chapter 11 to find out the critical value for rejection of the null hypothesis.

Our first task is to determine the degrees of freedom (*df*), which approximate the sample size. For this particular test statistic, the degrees of freedom are *n* – 1 where *n* equals the number of pairs of observations, or 25 – 1 = 24. These are the degrees of freedom for this test statistic only and not necessarily for any other.

Using this number (24), the level of risk you are willing to take (earlier defined as .05), and a one-tailed test (because there is a direction to the research hypothesis—the posttest score will be larger than the pretest score), the value needed for rejection of the null hypothesis is 1.711.

6. A comparison of the obtained value and the critical value is made.

The obtained value is 2.45, larger than the critical value needed for rejection of the null hypothesis.

7. and 8. Time for a decision.

Now comes our decision. If the obtained value is more extreme than the critical value, the null hypothesis cannot be accepted. If the obtained value does not exceed the critical value, the null hypothesis is the most attractive explanation. In this case, the obtained value does exceed the critical value—it is extreme enough for us to say that the difference between the pretest and the posttest did occur by something other than chance. And if we did our experiment correctly, then what could the factor be that affected the outcome? Easy—the introduction of the daily reading program. We know the difference is due to a particular factor. The difference between the pretest and the posttest groups could not have occurred by chance, but instead is due to the treatment.

So How Do I Interpret $t_{(24)}$ = 2.45, p < .05?

- *t* represents the test statistic that was used
- 24 is the number of degrees of freedom
- 2.45 is the obtained value using the formula we showed you earlier in the chapter

- $p < .05$ (the really important part of this little phrase) indicates that the probability is less than 5% on any one test of the null hypothesis that the average of posttest scores is greater than the average of pretest scores due to chance alone—there's something else going on. Because we defined .05 as our criterion for the research hypothesis being more attractive than the null hypothesis, our conclusion is that there is a significant difference between the two sets of scores. That's the something else.

USING THE COMPUTER TO PERFORM A T TEST

SPSS is willing and ready to help you perform these inferential tests. Here's how to perform the one that we just did and interpret the output. We are using the data set named Chapter 12 Data Set 1, which was also used in the earlier example.

1. Enter the data in the Data Editor. Be sure that there is a separate column for pretest and posttest scores. Unlike a *t* test for independent means, there are no groups to identify. In Figure 12.2, you can see how the cell entries are labeled pretest and posttest.

Figure 12.2   Data From Chapter 12 Data Set 1

Pretest	Posttest
3	7
5	8
4	6
6	7
5	8
5	9
4	6
5	6
3	7
6	8

2. Click Analyze → Compare Means → Paired-Samples T Test, and you will see the dialog box shown in Figure 12.3.

Figure 12.3   Paired-Samples *t* Test Dialog Box

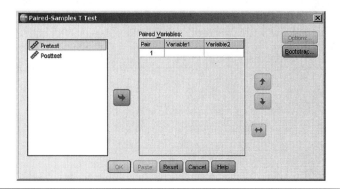

3. Drag the variable named Posttest to the Variable1 space in the Paired Variables: box as you see being done in Figure 12.4. Why drag Posttest first?

SPSS works in the following way. It subtracts Variable2 from Variable1 because the research hypothesis is nondirectional and "says" that the posttest scores will be greater than the prettest scores; thus, we want to subtract pretest scores from posttest scores and hence, posttest has to be identified as Variable1.

Figure 12.4   Dragging Variables for the Paired *t*-Test Analysis

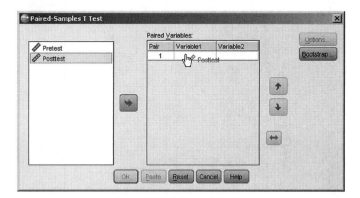

4. Drag the variable named Pretest to the Variable2 space in the Paired Variables: box.

5. Click OK.

6. SPSS will conduct the analysis and produce the output you see in Figure 12.5.

Figure 12.5 SPSS Output for a *t* Test Between Dependent Means

T-Test

Paired Samples Statistics

		Mean	N	Std. Deviation	Std. Error Mean
Pair 1	Pretest	6.32	25	1.725	.345
	Posttest	7.52	25	1.828	.366

Paired Samples Correlations

		N	Correlation	Sig.
Pair 1	Pretest & Posttest	25	.051	.810

Paired Samples Test

	Paired Differences					t	df	Sig. (2-tailed)
				95% Confidence Interval of the Difference				
	Mean	Std. Deviation	Std. Error Mean	Lower	Upper			
Pair 1 Pretest–Posttest	–1.200	2.449	.490	–2.211	–.189	–2.449	24	.022

> Earlier versions of SPSS (before version 19) do not allow you to
> define the order of which mean is subtracted from which other
> mean, as does version 18. So, if you are using a version of SPSS
> other than 18, then you will get a negative *t* value if the average of
> the first variable is smaller than the average of the second variable
> (which, in this case, would have been a *t* value of –2.449 instead
> of 2.449). As long as you keep the research hypothesis foremost in
> mind when you interpret the results, you should be fine.

What the SPSS Output Means

This SPSS output is rather straightforward. Here's a description of
the various components and once again, we're focusing only on
those that we think are within the scope of this book and most relevant to your understanding of the test we're discussing.

First, for both the pretest and the posttest, there are the reported
means, sample size, standard deviations, and the standard error of
the mean (a measure of sampling error). From this information, you
can immediately see that the posttest score (7.52) is larger than the
pretest score (6.32). At least this far into the analysis, it appears that
the results are supporting the research hypothesis that the children
scored higher on the posttest than the pretest.

Now for the results of interest—the actual values associated with
the *t* test. The difference between the means of the pretest and
posttest groups is 1.2, where the posttest mean was subtracted from
the pretest mean. And the exact probability that a *t* score of 2.449
was obtained by chance is .022—very unlikely.

But also notice that the output you see in Figure 12.5 shows this
to be the probability associated with a two-tailed (or nondirectional
test). We conducted a one-tailed test, so what to do? Read on!

Using Table B.2 in Appendix B, we find that for a one-tailed test,
with 24 degrees of freedom at the .05 level of significance, the critical value for rejection of the null hypothesis is 1.711 (see page
355). So, although SPSS will give us the specific *t* obtained value, it
will not give us the probability of that value for a one-tailed test. It
does for two-tailed, but not for one-tailed. For that, we have to rely
on our own skills and use the table as we did here, or else use a software program that can do one-tailed tests (see Chapter 19 for more
about this).

Believe it or not, way back in the olden days, when your author and perhaps your instructor were graduate students, there were only huge mainframe computers and not a hint of such marvels as we have today on our desktops. In other words, everything that was done in our statistics class was done only by hand. The great benefit of that is, first, it helps you to better understand the process. Second, should you be without a computer, you can still do the analysis. So, if the computer does not spit out all of what you need, use some creativity. As long as you know the basic formula for the critical value and have the appropriate tables, you'll do fine.

SUMMARY

That's it for means. You've just learned how to compare data from independent (Chapter 11) and dependent (Chapter 12) groups, and now it's time to move on to another class of significance tests that deals with more than two groups (be they independent or dependent). This class of techniques, called analysis of variance, is very powerful and popular and will be a valuable tool in your war chest!

TIME TO PRACTICE

1. What is the difference between a test of independent means and a test of dependent means, and when is each appropriate?

2. In the following examples, indicate whether you would perform a t test of independent means or dependent means.

    a. Two groups were exposed to different levels of treatment for ankle sprains. Which treatment was most effective?

    b. A researcher in nursing wanted to know if the recovery of patients was quicker when some received additional in-home care whereas others received the standard amount.

    c. A group of adolescent boys was offered interpersonal skills counseling and then tested in September and May to see if there was any impact on family harmony.

    d. One group of adult men was given instructions in reducing their high blood pressure whereas another was not given any instructions.

    e. One group of men was provided access to an exercise program and tested two times over a 6-month period for heart health.

3. For Chapter 12 Data Set 2, compute the t value manually and write a conclusion as to whether there was a change in tons of paper used as a function of the

recycling program in 25 different districts. (Hint: before and after become the two levels of treatment.) Test the hypothesis at the .01 level.

4. Here are the data from a study where adolescents were given counseling at the beginning of the school year to see if it had a positive impact on their tolerance for other adolescents who are ethnically different from them. Assessments were made right before the treatment and then 6 months later. Did the program work? The outcome variable is scored on an attitude toward others test with possible scores ranging from 0 to 50 and the higher the score, the more tolerance. Use SPSS or some other computer application to complete this analysis.

Before Treatment	After Treatment
45	46
46	44
32	47
34	42
33	45
21	32
23	36
41	43
27	24
38	41
41	38
47	31
41	22
32	36
22	36
34	27
36	41
19	44
23	32
22	32

5. For Chapter 12 Data Set 3, compute the *t* value and write a conclusion as to whether there is a difference in satisfaction level in a group of families' use of service centers following a social service intervention on a scale from 1 to 15. Do this exercise using SPSS, and report the exact probability of the outcome.

6. Do this exercise the good old-fashioned way, by hand. A famous brand-name manufacturer wants to know whether people prefer Nibbles or Wribbles. They get a chance to sample each type of cracker and indicate their like or dislike on a scale from 1 to 10. Which do they like the most?

Nibbles Rating	Wribbles Rating
9	4
3	7
1	6
6	8
5	7
7	7
8	8
3	6
10	7
3	8
5	9
2	8
9	7
6	3
2	6
5	7
8	6
1	5
6	5
3	6

13

Two Groups Too Many?

Try Analysis of Variance

Difficulty Scale ☺ (longer and harder than the others, but a very interesting and useful procedure—worth the work!)

WHAT YOU'LL LEARN ABOUT IN THIS CHAPTER

✦ What analysis of variance is and when it is appropriate to use

✦ How to compute and interpret the *F* statistic

✦ How to use SPSS to complete an analysis of variance

INTRODUCTION TO ANALYSIS OF VARIANCE

An increasingly popular area of psychology is the psychology of sports. Although the field focuses mostly on enhancing performance, many aspects of sports receive special attention. One aspect focuses on what psychological skills are necessary to be a successful athlete. With this question in mind, Marious Goudas, Yiannis Theodorakis, and Georgios Karamousalidis have tested the usefulness of the Athletic Coping Skills Inventory.

As part of their research, they used a simple **analysis of variance** (or ANOVA) to test the hypothesis that number of years of experience in sports is related to coping skill (or an athlete's score on the Athletic Coping Skills Inventory). ANOVA was used because more than two groups were being tested, and these groups were compared on their average performance. In particular, Group 1 included athletes with 6 years of experience or less, Group 2 included athletes with 7 to 10 years of experience, and Group 3 included athletes with more than 10 years of experience.

The test statistic for ANOVA is the *F* test (named for R. A. Fisher, the creator of the statistic), and the results showed that $F_{(2, 110)} =$ 13.08, $p < .01$. The means of the three groups did differ from one another in their score on the Peaking Under Pressure subscale of the test. In other words, any difference in test score is due to number of years of experience in athletics rather than some chance occurrence of scores.

Want to know more? Check out the original reference: Goudas, M., Theodorakis, Y., & Karamousalidis, G. (1998). Psychological skills in basketball: Preliminary study for development of a Greek form of the Athletic Coping Skills Inventory. *Perceptual and Motor Skills, 86,* 59–65.

The Path to Wisdom and Knowledge

Here's how you can use the flow chart shown in Figure 13.1 to select ANOVA as the appropriate test statistic. Follow along the highlighted sequence of steps.

1. We are testing for differences between scores of the different groups, in this case, the difference between the peaking scores of athletes.

2. The athletes are not being tested more than once.

3. There are three groups (less than 6 years, 7–10 years, and more than 10 years of experience).

4. The appropriate test statistic is simple analysis of variance.

Different Flavors of ANOVA

ANOVA comes in many different flavors. The simplest kind, and the focus of this chapter, is the **simple analysis of variance**, where there is one factor or one treatment variable (such as group membership) being explored, and there are more than two levels within this factor. Simple ANOVA is also called **one-way analysis of variance** because there is only one grouping dimension. The technique is called analysis of variance because the variance due to differences in performance is separated into variance that's due to differences between individuals *within* groups and variance due

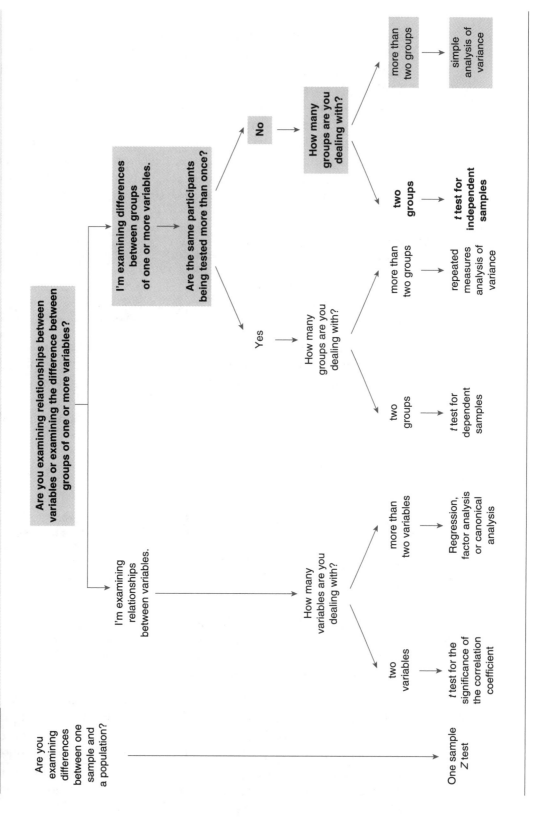

to differences *between* groups. Then, the two types of variance are compared with one another.

In fact, ANOVA is, in many ways, similar to a *t* test. In both procedures, differences between means are computed. But with ANOVA, there are more than two means.

For example, let's say we were investigating the effects on language development of being in preschool for 5, 10, or 20 hours per week. The group to which the children belong is the treatment variable, or the grouping factor. Language development is the outcome measure. The experimental design looks something like this, with three levels of the one variable (hours participating).

Group 1 (5 hours per week)	Group 2 (10 hours per week)	Group 3 (20 hours per week)
Language development test score	Language development test score	Language development test score

The more complex type of ANOVA is called a **factorial design,** where there is more than one treatment factor being explored. Here's an example where the effect of number of hours of preschool participation is being examined, but the effects of gender differences are being examined as well. The experimental design can look something like this:

	Number of Hours of Preschool Participation		
Gender	Group 1 (5 hours per week)	Group 2 (10 hours per week)	Group 3 (20 hours per week)
Male	Language development test score	Language development test score	Language development test score
Female	Language development test score	Language development test score	Language development test score

This factorial design is described as a 3 × 2 factorial design. The 3 indicates that there are three levels of one grouping factor (Group 1, Group 2, and Group 3). The 2 indicates that there are two levels of the other grouping factor (male and female). In combination, there are 6 different possibilities (males who spend 5 hours per week in preschool, females who spend 5 hours per week in preschool, males who spend 10 hours per week in preschool, etc.).

These factorial designs follow the same basic logic and principles of simple ANOVA, but they are just more ambitious in that they can test the influence of more than one factor at a time as well as a combination of factors. Don't worry—you'll learn all about factorial designs in the next chapter.

COMPUTING THE F TEST STATISTIC

Simple ANOVA involves testing the difference between the means of more than two groups on one factor or dimension. For example, you might want to know whether four groups of people (20, 25, 30, and 35 years of age) differ in their attitude toward public support of private schools. Or, you might be interested in determining whether five groups of children from different grades (2nd, 4th, 6th, 8th, and 10th) differ in the level of parental participation in school activities.

Any analysis where

- there is only one dimension or treatment,
- there are more than two levels of the grouping factor, and
- one is looking at differences across groups in average scores

requires that simple ANOVA be used.

The formula for the computation of the F value, which is the test statistic needed to evaluate the hypothesis that there are overall differences between groups, is shown in Formula 13.1. It is simple at this level, but it takes a bit more effort to compute than some of the other test statistics you have worked with in earlier chapters.

$$F = \frac{MS_{\text{between}}}{MS_{\text{within}}}$$

(13.1)

Tech Talk

The logic behind this ratio goes something like this. If there was absolutely no variability within each group (all the scores were the same), then any difference between groups would be meaningful, right? Probably so. The ANOVA formula (which is a ratio) compares the amount of variability between groups (which is due to the grouping factor) to the amount of variability within groups (which is due to chance). If that ratio is 1, then the amount of variability due to within-group differences is equal to the amount of variability due to between-group differences, and any difference between groups would not be significant. As the average difference between groups gets larger (and the numerator of the ratio increases in value), the F value increases as well. As the F value increases, it becomes more extreme in relation to the distribution of all F values and is more likely due to something other than chance. Whew!

Here are some data and some preliminary calculations to illustrate how the F value is computed. For our example, let's assume these are three groups of preschoolers and their language scores.

Group 1 Scores	Group 2 Scores	Group 3 Scores
87	87	89
86	85	91
76	99	96
56	85	87
78	79	89
98	81	90
77	82	89
66	78	96
75	85	96
67	91	93

Here are the famous eight steps and the computation of the F test statistic.

1. A statement of the null and research hypotheses.

The null hypothesis, shown in Formula 13.2, states that there is no difference between the means for the three different groups. ANOVA, also called the F test (because it produces an F statistic or an F ratio), looks for an overall difference between groups.

It does not look at pairwise differences, such as the difference between Group 1 and Group 2. For that, we have to use another technique, which we will discuss later in the chapter.

$$H_0: \mu_1 = \mu_2 = \mu_3 \qquad (13.2)$$

The research hypothesis, shown in Formula 13.3, states that there is an overall difference between the means of the three groups. Note that there is no direction to the difference because all F tests are nondirectional.

$$H_1 : \overline{X}_1 \neq \overline{X}_2 \neq \overline{X}_3 \qquad\qquad (13.3)$$

> Up to now, we've talked about one- and two-tailed tests. No such thing when talking about ANOVA. Because more than two groups are being tested, and because the F test is an omnibus (how's that for a word?) test (meaning that it tests for an overall difference between means), talking about the direction of specific differences does not make any sense.

2. Setting the level of risk (or the level of significance or Type I error) associated with the null hypothesis.

The level of risk or Type I error or level of significance (any other names?) is .05. Once again, the level of significance used is totally at the discretion of the researcher.

3. Selection of the appropriate test statistic.

Using the flow chart shown in Figure 13.1, we determined that the appropriate test is a simple ANOVA.

4. Computation of the test statistic value (called the obtained value).

Now's your chance to plug in values and do some computation. There's a good deal of computation to do.

- The F ratio is a ratio of variability between groups to variability within groups. To compute these values, we first have to compute what is called the sum of squares for each source of variability—between groups, within groups, and the total.
- The between-group sum of squares is equal to the sum of the differences between the mean of all scores and the mean of each group's score, which is then squared. This gives us an idea of how different each group's mean is from the overall mean.
- The within-group sum of squares is equal to the sum of the differences between each individual score in a group and the

mean of each group, which is then squared. This gives us an idea how different each score in a group is from the mean of that group.

- The total sum of squares is equal to the sum of the between-group and within-group sum of squares. OK—let's figure these values.

Here are the practice data you saw previously with all the calculations you need to compute the between-group, within-group, and total sum of squares.

First, let's look at what we have in this expanded table. Starting down the left of the table (see Figure 13.2):

n	is the number of participants in each group (such as 10)
ΣX	is the sum of the scores in each group (such as 766)
$\overline{X}$	is the mean of each group (such as 76.60)
$\Sigma(X^2)$	is the sum of each score squared (such as 59,964)
$(\Sigma X)^2/n$	is the sum of the scores in each group squared and then divided by the size of the group (such as 58,675.60)

Second, let's look at the right-most column:

N	is the total number of participants (such as 30)
$\Sigma\Sigma X$	is the sum of all the scores across groups
$(\Sigma\Sigma X)^2/N$	is the sum of all the scores across groups squared and divided by N
$\Sigma\Sigma(X^2)$	is the sum of all the sums of squared scores
$\Sigma(\Sigma X)^2/n$	is the sum of the sum of each group's scores squared divided by n

That is a load of computation to carry out, and we are almost finished.

First, we compute the sum of squares for each source of variability. Here are the calculations:

Between sum of squares	$\Sigma(\Sigma X)^2/n - (\Sigma\Sigma X)^2/N$ or $215,171.60 - 214,038.53$	1,133.07
Within sum of squares	$\Sigma\Sigma(X^2) - \Sigma(\Sigma X)^2/n$ or $216,910 - 215,171.6$	1,738.40
Total sum of squares	$\Sigma\Sigma(X^2) - (\Sigma\Sigma X)^2/N$ or $216,910 - 214,038.53$	2,871.47

Figure 13.2 Computing the Important Values for a One-Way ANOVA

Group	Test Score	X^2	Group	Test Score	X^2	Group	Test Score	X^2
1	87	7,569	2	87	7,569	3	89	7,921
1	86	7,396	2	85	7,225	3	91	8,281
1	76	5,776	2	99	9,801	3	96	9,216
1	56	3,136	2	85	7,225	3	87	7,569
1	78	6,084	2	79	6,241	3	89	7,921
1	98	9,604	2	81	6,561	3	90	8,100
1	77	5,929	2	82	6,724	3	89	7,921
1	66	4,356	2	78	6,084	3	96	9,216
1	75	5,625	2	85	7,225	3	96	9,216
1	67	4,489	2	91	8,281	3	93	8,649
n	10			10			10	
ΣX	766			852			916	
$\overline{X}$	76.60			85.20			91.60	
$\Sigma(X^2)$	59,964			72,936			84,010	
$(\Sigma X)^2/n$	58,675.60			72,590.40			83,905.60	

$N = 30.00$

$\Sigma\Sigma X = 2{,}534.00$

$(\Sigma\Sigma X)^2/N = \mathbf{214{,}038.53}$

$\Sigma\Sigma(X^2) = \mathbf{216{,}910}$

$\Sigma(\Sigma X)^2/n = \mathbf{215{,}171.60}$

Second, we need to compute the mean sum of squares, which is simply an average sum of squares. These are the variance estimates that we need to eventually compute the all-important F ratio.

We do that by dividing each sum of squares by the appropriate number of degrees of freedom (df). Remember, degrees of freedom are an approximation of the sample or group size. We need two sets of degrees of freedom for ANOVA. For the between-group estimate, it is $k - 1$, where k equals the number of groups (in this case, there are 3 groups and 2 degrees of freedom), and for the within-group estimate, we need $N - k$, where N equals the total sample size (which means that the number of degrees of freedom is $30 - 3$, or 27). And the F ratio is simply a ratio of the mean sum of squares due to between-group differences over the mean sum of squares due to within-group differences, or $566.54/64.39 = 8.799$. This is the obtained F value.

Here's a summary table of the variance estimates used to compute the F ratio and how most F tables appear in professional journals and manuscripts.

Source	Sum of Squares	df	Mean Sum of Squares	F
Between groups	1,133.07	2	566.54	8.799
Within groups	1,738.40	27	64.39	
Total	2,871.47	29		

All that trouble for one little F ratio. But as we have said earlier, it's essential to do these procedures at least once by hand. It gives you the important appreciation of where the numbers come from and some insight into what they mean.

Because you already know about t tests, you might be wondering how a t value (which is always used for the test between the difference of the means for two groups) and an F value (which is always used for more than two groups) might be related. Interestingly enough, an F value for two groups is equal to a t value for two groups squared, or $F = t^2$. Handy trivia question, right? But also useful if you know one and need to know the other.

5. Determination of the value needed for rejection of the null hypothesis using the appropriate table of critical values for the particular statistic.

As we have done before, we have to compare the obtained and critical values. We now need to turn to the table that lists the critical values for the F test, Table B.3 in Appendix B. Our first task is to determine the degrees of freedom for the numerator, which is $k - 1$, or $3 - 1 = 2$. Then determine the degrees of freedom for the denominator, which is $N - k$, or $30 - 3 = 27$. Together, they are represented as $F_{(2, 27)}$.

The obtained value is 8.80, or $F_{(2, 27)} = 8.80$. The critical value at the .05 level with 2 degrees of freedom in the numerator (represented by columns in Table B.3) and 27 degrees of freedom in the denominator (represented by rows in Table B.3) is 3.36. So at the .05 level, with 2 and 27 degrees of freedom for an omnibus test between the means of the three groups, the value needed for rejection of the null hypothesis is 3.36.

6. A comparison of the obtained value and the critical value is made.

The obtained value is 8.80, and the critical value for rejection of the null hypothesis at the .05 level that the three groups are different from one another (without concern for where the difference lies) is 3.36.

7. and 8. Decision time.

Now comes our decision. If the obtained value is more extreme than the critical value, the null hypothesis cannot be accepted. If the obtained value does not exceed the critical value, the null hypothesis is the most attractive explanation. In this case, the obtained value does exceed the critical value—it is extreme enough for us to say that the difference between the three groups is not due to chance. And if we did our experiment correctly, then what could the factor be that affected the outcome? Easy—the number of hours of preschool. We know the difference is due to a particular factor because the difference between the groups could not have occurred by chance, but instead is due to the treatment.

So How Do I Interpret $F_{(2, 27)} = 8.80$, $p < .05$?

- F represents the test statistic that was used
- 2, 27 are the numbers of degrees of freedom for the between-group and within-group estimates
- 8.80 is the obtained value using the formula we showed you earlier in the chapter

- $p < .05$ (the really important part of this little phrase) indicates that the probability is less than 5% on any one test of the null hypothesis that the average scores of each group's language skills differ due to chance alone rather than the effect of the treatment. Because we defined .05 as our criterion for the research hypothesis being more attractive than the null hypothesis, our conclusion is that there is a significant difference between the three sets of scores.

(Really Important) Tech Talk

Imagine this scenario. You're a high-powered researcher at an advertising company, and you want to see if color makes a difference in sales. And you'll test this at the .05 level. So you put together a brochure that is all black and white, one that is 25% color, the next 50%, then 75%, and finally, 100% color, for five different levels. You do an ANOVA and find out that there is a difference. But because ANOVA is an omnibus test, you don't know where the source of the significant difference lies. So you take two groups (or pairs) at a time (such as 25% color and 75% color) and test them against each other. In fact, you test every combination of 2 against each other. Kosher? No way. This is called performing multiple t tests, and it is actually against the law in some jurisdictions. When you do this, the Type I error rate (which you set at .05) balloons depending on the number of tests you want to conduct. There are 10 possible comparisons (no color vs. 25%, no color vs. 50%, no color vs. 75%, etc.), and the real Type I error rate is $1 - (1 - \alpha)^k$, where

α is the Type I error rate, which is .05 in this example

k is the number of comparisons

So, instead of .05, the actual error rate that each comparison is being tested at is

$$1 - (1 - .05)^{10} = .40 \ (!!!!!)$$

Surely not .05. Quite a difference, no?

USING THE COMPUTER TO COMPUTE THE F RATIO

The F ratio is not an easy value to compute by hand. That's all there is to it. Using the computer is much easier and more accurate because it eliminates any computational errors. That said, you should be glad you have seen the value computed manually because it's an important skill to have and helps you understand the concepts behind the process. But also be glad that there are tools such as SPSS.

We'll use the data found in Chapter 13 Data Set 1, which was used in the above preschool example.

1. Enter the data in the Data Editor. Be sure that there is a column for group and that you have three groups represented in that column. In Figure 13.3, you can see how the cell entries are labeled Group and Language_Score

Figure 13.3 Data From Chapter 13 Data Set 1

Group	Language_Score
5 Hours	87
5 Hours	86
5 Hours	76
5 Hours	56
5 Hours	78
5 Hours	98
5 Hours	77
5 Hours	66
5 Hours	75
5 Hours	67

2. Click Analyze → Compare Means → One-Way ANOVA and you will see the One-Way ANOVA dialog box shown in Figure 13.4.

Figure 13.4 One-Way ANOVA Dialog Box

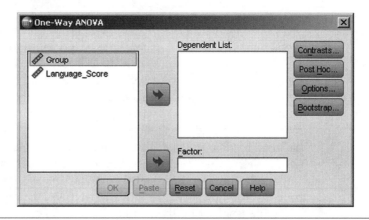

3. Click on the variable named Group, and then click to move it to the Factor box.

4. Click on the variable named Language_Score, and then click ⤵ to move it to the Dependent List: box.

5. Click Options and then Descriptives and then Continue.

6. Click OK. SPSS will conduct the analysis and produce the output you see in Figure 13.5.

What the SPSS Output Means

This SPSS output is straightforward and looks just like the table that we created earlier to show you how to compute the F ratio along with some descriptive statistics. Here's what we have.

1. The source of the variance as between-group, within-group, and total is identified.

2. Next, we have the respective sum of squares for each source.

3. The degrees of freedom follow, then followed by the mean square, which is the sum of squares divided by the degrees of freedom.

4. Finally, there's the obtained value and the associated level of significance.

Keep in mind that this hypothesis was tested at the .05 level. The SPSS output provides the exact probability of the outcome, .001— much more accurate and much more unlikely than .05.

Tech Talk

OK, so you've run an ANOVA and you know that there is an overall difference between the means of three or four or more groups. But where does that difference lie? You already know not to perform multiple t tests. You need to perform what are called **post hoc,** or after-the-fact, comparisons. Here's where each mean is compared to each other mean and you can see where the difference lies, but what's most important is that the Type I error for each comparison is controlled at the same level as you set. There are a bunch of these different comparisons, among them being the Bonferroni (your dear author's favorite statistical term). To complete this specific analysis using SPSS, you click the Post Hoc option you see in the ANOVA dialog box (Figure 13.4), then click Bonferroni, then Continue, and so on, and you'll see output

Figure 13.5 SPSS Output for a One-Way Analysis of Variance

	N	Mean	Std. Deviation	Std. Error	95% Confidence Interval for Mean		Minimum	Maximum
					Lower Bound	Upper Bound		
5 Hours	10	76.60	11.965	3.784	68.04	85.16	56	98
10 Hours	10	85.20	6.197	1.960	80.77	89.63	78	99
20 Hours	10	91.60	3.406	1.077	89.16	94.04	87	96
Total	30	84.47	9.951	1.817	80.75	88.18	56	99

ANOVA

Language_Score

	Sum of Squares	df	Mean Square	F	Sig.
Between Groups	1133.067	2	566.533	8.799	.001
Within Groups	1738.400	27	64.385		
Total	2871.467	29			

Figure 13.6 Post Hoc Comparisons After a One-Way ANOVA

Post Hoc Tests

Multiple Comparisons

Dependent Variable: Language_Score
Bonferroni

(I) Group	(J) Group	Mean Difference (I-J)	Std. Error	Sig.	95% Confidence Interval	
					Lower Bound	Upper Bound
5 Hours	10 Hours	-8.600	3.588	.071	-17.76	.56
	20 Hours	-15.000*	3.588	.001	-24.16	-5.84
10 Hours	5 Hours	8.600	3.588	.071	-.56	17.76
	20 Hours	-6.400	3.588	.257	-15.56	2.76
20 Hours	5 Hours	15.000*	3.588	.001	5.84	24.16
	10 Hours	6.400	3.588	.257	-2.76	15.56

*The mean difference is significant at the .05 level.

something like that shown in Figure 13.6. It's really simple to see how this analysis tells you that the significant pairwise differences between the groups contributing to the overall significant difference between all three groups lies between Groups 1 and 3 and that there is no pairwise difference between Groups 1 and 2 or Groups 2 and 3. This pairwise stuff is very important because it allows you to understand the source of the difference between more than two groups.

SUMMARY

Analysis of variance (ANOVA) is the most complex of all the inferential tests you will learn in *Statistics for People Who (Think They) Hate Statistics*. It takes a good deal of concentration to perform the manual calculations, and even when you use SPSS, you have to be on your toes to understand that this is an overall test, and one part will not give you information about differences between pairs of treatments. If you choose to go on and do post hoc analysis, you're really completing all the tasks that go along with the powerful tool. Only one more test between averages and that's a factorial ANOVA. This is the Holy Grail of ANOVAs and can involve two or more factors, but we'll stick with two and SPSS will show us the way.

TIME TO PRACTICE

1. When is analysis of variance a more appropriate use of a statistical technique than a test between a pair of means?

2. Using the following table, provide three examples of a simple one-way ANOVA, two examples of a two-factor ANOVA, and one example of a three-factor ANOVA. We show you some examples, you show the others. Be sure to identify the grouping and the test variable as we have done here.

Design	Grouping Variable(s)	Test Variable
Simple ANOVA	Four levels of hours of training— 2, 4, 6, and 8 hours	Typing accuracy
	Enter Your Example Here	Enter Your Example Here
	Enter Your Example Here	Enter Your Example Here
	Enter Your Example Here	Enter Your Example Here
Two-factor ANOVA	Two levels of training and gender (2 × 2 design)	Typing accuracy

(Continued)

(Continued)

Design	Grouping Variable(s)	Test Variable
	Enter Your Example Here	Enter Your Example Here
	Enter Your Example Here	Enter Your Example Here
Three-factor ANOVA	Two levels of training and two of gender and three of income	Voting attitudes
	Enter Your Example Here	Enter Your Example Here

3. Using the data in Chapter 13 Data Set 2 and SPSS, compute the F ratio for a comparison between the three levels representing the average amount of time that swimmers practice weekly (<15, 15–25, and >25 hours) with the outcome variable being their time for the 100-yard freestyle. Answer the question whether practice time makes a difference. Don't forget to use the Options feature to get the means for the groups.

4. The following data were collected by a researcher who wanted to know if the amount of stress is different for three groups of employees and can be found in Chapter 13, Data Set 3. Group 1 employees work the morning/day shift, Group 2 employees work the day/evening shift, and Group 3 employees work the night shift. The null hypothesis is that there is no difference in the amount of stress between groups. Test this in SPSS and provide your conclusion.

14

Two Too Many Factors

Factorial Analysis of Variance: A Brief Introduction

Difficulty Scale ☺ (about as tough as they get for the challenging ideas—but we're only touching on the main concepts here)

WHAT YOU'LL LEARN ABOUT IN THIS CHAPTER

✦ When to use analysis of variance with more than one factor

✦ All about main and interaction effects

✦ How to use SPSS to complete a factorial analysis of variance

INTRODUCTION TO FACTORIAL ANALYSIS OF VARIANCE

How people make decisions has been one of the processes that have fascinated psychologists for decades. The data that have resulted from the studies have been applied to such broad fields as advertising, business, planning, and even theology. Miltiades Proios and George Doganis investigated the effect of how the experience of being actively involved in the decision-making process (in a variety of settings) and age can have an impact on moral reasoning. The sample consisted of a total of 148 referees—56 who referee soccer, 55 who referee basketball, and 37 who referee handball. Their ages ranged from 17 to 50 years, and gender was not considered an important variable. Within the entire sample, about 8% had not had any experience in social, political, or athletic settings where they fully participated in the decision-making process, about 53% were active but not fully participatory, and about 39% were both active and did participate in the decisions made within that organization. A two-way (multivariate—see Chapter 18 for more about this) analysis of variance showed an interaction between experience and age on moral reasoning and goal orientation of referees.

Why a two-way analysis of variance? Easy—there were two independent factors, with the first being level of experience and the second being age. Here, just as with any analysis of variance procedure, there is

1. a test of the **main effect** for age,

2. a test of the main effect for experience, and

3. a test for the interaction between experience and age (which turned out to be significant).

The very cool thing about analysis of variance when more than one factor or independent variable is tested is that the researcher can look at the individual effects of each factor, but also the simultaneous effects of both, through what is called an interaction, which we will talk about more later in this chapter.

Want to know more? Proios, M., & Doganis, G. (2003). Experiences from active membership and participation in decision-making processes and age in moral reasoning and goal orientation of referees. *Perceptual and Motor Skills, 96*(1), 113–126.

The Path to Wisdom and Knowledge

Here's how you can use the flow chart shown in Figure 14.1 to select ANOVA (but this time with more than one factor) as the appropriate test statistic. Follow along the highlighted sequence of steps.

As in Chapter 13, we have already decided that ANOVA is the correct procedure (examining differences in more than two groups or levels of the independent variable), but because we have more than one factor, factorial ANOVA is the right choice.

1. We are testing for differences between scores of the different groups, in this case, the difference between the level of experience and age.

2. The participants are not being tested more than once.

3. We are dealing with more than two groups.

4. We are dealing with more than one factor or independent variable.

5. The appropriate test statistic is factorial analysis of variance.

Figure 14.1 Determining That Factorial Analysis of Variance Is the Correct Test Statistic

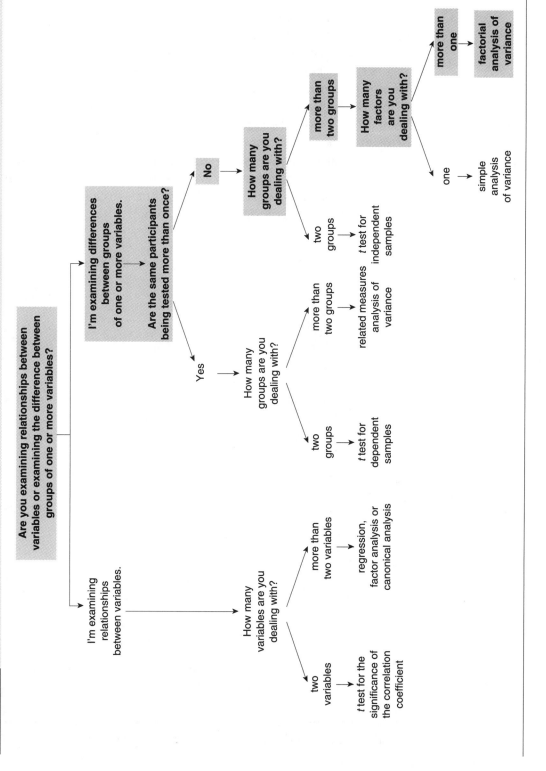

A New Flavor of ANOVA

You know that ANOVA comes in at least one flavor, the simple analysis of variance we discussed in Chapter 13. There is one factor or one treatment variable (such as group membership) being explored, and there are more than two groups or levels within this factor or treatment variable.

Now, we bump up the entire technique a notch to include the exploration of more than one factor simultaneously and call this a **factorial analysis of variance.**

Let's look at a simple example that includes two factors, gender (male or female) and treatment, which is type of exercise program (high impact or low impact) and the outcome—weight loss. Here's what the experimental design would look like:

		Exercise Program	
		High Impact	Low Impact
Gender	Male		
	Female		

Then, we will look at what main effects and an interaction look like. Not a lot of data analysis here until a bit later on in the chapter—mostly just look and learn. Even though the exact probabilities of a Type I error are provided in the results (and we don't have to mess with statements like $p < .05$ and such), we'll use .05 as the criterion for rejection of the null.

There are three questions that you can ask and answer from this type of analysis.

1. Is there a difference between the effects on weight loss between two levels of exercise program, high impact and low impact?

2. Is there a difference between the effects on weight loss between two levels of gender, male and female?

3. Is the effect of being in the high- or low-impact program different for males or females (this is the interaction question)?

Questions 1 and 2 deal with the presence of main effects, whereas Question 3 deals with the interaction between the two factors.

THE MAIN EVENT: MAIN EFFECTS IN FACTORIAL ANOVA

You might remember that the primary task of analysis of variance is to test for the difference between two or more groups. When an analysis of the data reveals a difference between the levels of any factor, we talk about there being a **main effect.** Here's an example where there are 10 participants in each of the four groups in the above example, for a total of 40. And, here's what the results of the analysis look like (and we used SPSS to compute this fancy pants table). This is called a **source table.**

Tests of Between-Subjects Effects

Dependent Variable: LOSS

Source	Type III Sum of Squares	df	Mean Square	F	Sig.
Corrected Model	3678.275	3	1226.092	8.605	.000
Intercept	232715.025	1	232715.025	1633.183	.000
TREATMEN	429.025	1	429.025	3.011	.091
GENDER	3222.025	1	3222.025	22.612	.000
TREATMEN *					
GENDER	27.225	1	27.225	.191	.665
Error	5129.700	36	142.492		
Total	241523.000	40			
Corrected Total	8807.975	39			

Pay attention to only the *Source* and the *Sig.* columns (which are highlighted). The conclusion we can reach is that there is a main effect for gender (p = .000), no main effect for treatment (p = .091), and no interaction between the two main factors (p = .665). So, as far as weight loss, it didn't matter whether one was in the high- or low-impact group, but it did matter if one were male or female. And because there was no interaction between the treatment factor and gender, there were no differential effects for treatment across gender.

If you plotted the means of these values, you would get something that looks like Figure 14.2:

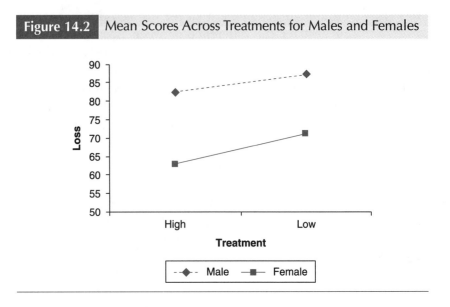

Figure 14.2 Mean Scores Across Treatments for Males and Females

You can see a big difference in distance on the loss axis between males and females (average score for all males is 85.25 and for females is 67.30), but for treatment (if you computed the averages), you would find there to be little difference (with the average score across all highs being 73.00 and across all lows being 79.55). Now, of course, this is an analysis of variance, and the variability in the groups does matter, but in this example, you can see the differences between groups (such as males and females) within each factor (such as gender) and how they are reflected by the results of the analysis.

EVEN MORE INTERESTING: INTERACTION EFFECTS

OK—now let's move to the interaction. Let's look at a new source table that indicates men and women are affected differentially across treatments, indicating the presence of an **interaction effect**. And, indeed, you will see some very cool outcomes. Once again, most important parts are bolded.

Tests of Between-Subjects Effects

Dependent Variable: LOSS

Source	Type III Sum of Squares	df	Mean Square	F	Sig.
Corrected Model	1522.875a	3	507.625	4.678	.007
Intercept	218892.025	1	218892.025	2017.386	.000
TREATMEN	265.225	1	265.225	2.444	**.127**
GENDER	207.025	1	207.025	1.908	**.176**
TREATMEN * GENDER	1050.625	1	1050.625	9.683	**.004**
Error	3906.100	36	108.503		
Total	224321.000	40			
Corrected Total	5428.975	39			

Here, there is no main effect for treatment or gender ($p = .127$ and .176, respectively), but yikes, there is one for the treatment by gender interaction ($p = .004$), which makes this a very interesting outcome. In effect, it does not matter if you are in the high- or low-impact treatment group, or if you are male or female, but it does matter a whole lot if you are in both conditions simultaneously such that the treatment does have an impact differentially on the weight loss of males than on females.

Figure 14.3 shows what a chart of the averages for each of the four groups looks like.

And, here's what the actual means themselves look like (all compliments of SPSS):

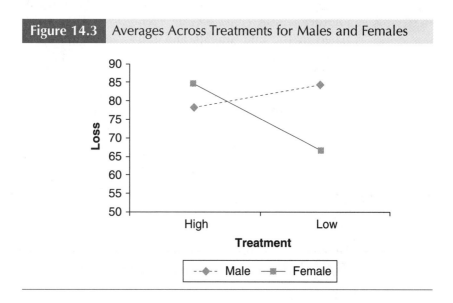

Figure 14.3 Averages Across Treatments for Males and Females

	Male	Female
High Impact	73.70	79.40
Low Impact	78.80	64.00

What to make of this? Well, the interpretation here is pretty straightforward. Here's what we can say, and, being as smart as you are, you can recognize that these are the answers to the three questions we listed earlier.

1. There is no main effect for type of exercise.

2. There is no main effect for gender.

3. There is a clear interaction between treatment and gender, which means females lose more weight than males under the high-impact treatment condition, and males lose more weight than females under the low-impact condition.

THINGS TO REMEMBER

This is all pretty remarkable stuff. If you didn't know any better (and never read this chapter), you would think that all you have to do is a simple *t* test between the averages for males and females, and then another simple *t* test for the averages between those who participated in the high-impact and those who participated in the low-impact treatment—and you would have found nothing. But, using the idea of an interaction between main factors, you find out that there is a differential effect—an outcome that would have gone unnoticed otherwise. Indeed, if you can bear the admission, interactions really are the most interesting outcomes in any factorial analysis of variance.

COMPUTING THE TEST STATISTIC

Here's a change for you. Throughout *Statistics for People Who (Think They) Hate Statistics*, we have provided you with examples of how to perform particular techniques the old-fashioned way (by hand, using a calculator) as well as using a statistical analysis package such as SPSS. With the introduction of factorial ANOVA, we are illustrating the analysis using only SPSS—not that it is any more of an intellectual challenge to complete a factorial ANOVA using a calculator, but it certainly is more laborious. It's for that reason we are not going to cover the computation by hand, but go right to the computation of the important values and spend more time on the interpretation.

We'll use the data that show there is a significant interaction, as seen here.

Treatment →	High Impact	High Impact	Low Impact	Low Impact
Gender→	Male	Female	Male	Female
	76	65	88	65
	78	90	76	67
	76	65	76	67
	76	90	76	87
	76	65	56	78
	74	90	76	56
	74	90	76	54
	76	79	98	56
	76	70	88	54
	55	90	78	56

Here are the steps and the computation of the F test statistic. The reason why you don't see the "famous eight steps" (so OK, there are 10) is that this is the first (and only) time throughout the book that we don't do the computations by hand, but use only the computer. The analysis (as I said before) is just too labor intensive for a course at this level.

1. A statement of the null and research hypotheses.

There are actually three null hypotheses, shown here (Formulas 14.1a, 14.1b, and 14.1c) that state that there is no difference between the means for the two factors, and no interaction. Here we go.

First for the treatment,

$$H_0: \mu_{high} = \mu_{low} \tag{14.1a}$$

and now for gender,

$$H_0: \mu_{male} = \mu_{female} \tag{14.1b}$$

and now for the interaction between treatment and gender,

$$H_0: \mu_{high \bullet male} = \mu_{high \bullet female} = \mu_{low \bullet male} = \mu_{low \bullet female} \tag{14.1c}$$

The research hypotheses, shown in Formulas 14.2a, 14.2b, and 14.2c, state that there is a difference between the means of the groups, and there is an interaction. Here they are.

First for the treatment,

$$H_1: \bar{X}_{high} \neq \bar{X}_{low} \tag{14.2a}$$

and now for gender,

$$H_1: \bar{X}_{male} \neq \bar{X}_{female} \tag{14.2b}$$

and now for the interaction between treatment and gender,

$$H_1: \bar{X}_{high \bullet male} \neq \bar{X}_{high \bullet female} \quad \bar{X} \neq_{low \bullet male} \quad \bar{X} \neq_{low \bullet female} \tag{14.2c}$$

2. Setting the level of risk (or the level of significance or Type I error) associated with the null hypothesis.

The level of risk or Type I error or level of significance is .05. Once again, the SPSS analysis will provide the exact p value, but the level of significance used is totally at the discretion of the researcher.

3. Selection of the appropriate test statistic.

Using the flow chart shown in Figure 14.1, we determined that the appropriate test is a factorial ANOVA.

4. Computation of the test statistic value (called the obtained value).

We'll use SPSS for this, and here are the steps. We'll use the above data, which are available on the website for the book (http:// www.sagepub.com/salkind4e and www.onlinefilefolder.com—see the *Statistics for People* . . . introduction for complete information on downloading files) as Chapter 14 Data Set 1 (and are listed in Appendix C).

Enter the data in the Data Editor or open the file. Be sure that there is a column for each factor, including treatment, gender, and loss, as you see in Figure 14.4.

Figure 14.4 Data From Chapter 14 Data Set 1

Treatment	Gender	Loss
High Impact	Male	76
High Impact	Male	78
High Impact	Male	76
High Impact	Male	76
High Impact	Male	76
High Impact	Male	74
High Impact	Male	74
High Impact	Male	76
High Impact	Male	76
High Impact	Male	55

---

5. Click Analyze → General Linear Model → Univariate and you will see the Factorial ANOVA dialog box shown in Figure 14.5.

Figure 14.5 Factorial ANOVA Dialog Box

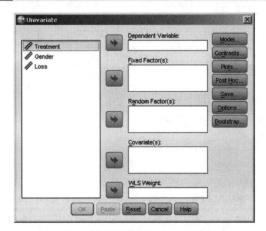

6. Click on the variable named Loss and then click ⬦ to move it to the Dependent Variable: box.

7. Click on the variable named Treatment and then click ⬦ to move it to the Fixed Factor(s): box.

8. Click on the variable named Gender and then click ⬦ to move it to the Random Factor(s): box.

9. Click Options and then Descriptive Statistics and then click Continue.

10. Click OK. SPSS will conduct the analysis and produce the output you see in Figure 14.6 (as you saw earlier in the chapter).

| Figure 14.6 | SPSS Output for a Factorial Analysis of Variance |

Univariate Analysis of Variance

Between-Subjects Factors

		Value Label	N
Treatment	1	High Impact	20
	2	Low Impact	20
Gender	1	Male	20
	2	Female	20

Test of Between-Subjects Effects

Dependent Variable: Loss

Source		Type III Sum of Squares	df	Mean Square	F	Sig.
Intercept	Hypothesis	218892.025	1	218892.025	1057.322	.020
	Error	207.025	1	207.025[a]		
Treatment	Hypothesis	265.225	1	265.225	.252	.704
	Error	1050.625	1	1050.625[b]		
Gender	Hypothesis	207.025	1	207.025	.197	.734
	Error	1050.625	1	1050.625[b]		
Treatment	Hypothesis	1050.625	1	1050.625	9.683	.004
* Gender	Error	3906.100	36	108.503[c]		

a. MS(Gender)

b. MS(Treatment * Gender)

c. MS(Error)

Wondering why the SPSS output is labeled *Univariate* Analysis of Variance? We knew you were. Well, in SPSS talk, this is an analysis that looks at only one dependent or outcome variable—in this case, weight loss. If we had more than one variable as part of the research question (such as attitude toward eating), then it would be a multivariate analysis of variance, which looks at group differences and more than one dependent variable, but also controls for the relationship between the dependent variables. More about this in Chapter 18.

What the SPSS Output Means

This SPSS output is pretty straightforward. Here's what we have.

1. The source of the variance as between-group, within-group, and total is identified.

2. Next, we have the respective sum of squares for each source.

3. The degrees of freedom follow, and then the mean square, which is the sum of squares divided by the degrees of freedom.

4. Finally, there's the obtained value and the exact level of significance.

5. For gender, the results, as printed in a journal article or report, would look something like $F_{(1,36)} = .197$, $p = .734$.

6. For the treatment, the results, as printed in a journal article or report, would look something like $F_{(1,36)} = .252$, $p = .704$.

7. And for the interaction, the results, as printed in a journal article or report, would look something like $F_{(1,36)} = 9.683$, $p = .004$.

We're done!

SUMMARY

Now that we are done, done, done with testing differences between means, we'll move on to examining the significance of correlations, or the relationship between two variables.

TIME TO PRACTICE

1. When would you use a factorial ANOVA rather than a simple ANOVA to test the significance of the difference between the average of two or more groups?

2. Create a drawing or plan for a 2 × 3 experimental design that would lend itself to a factorial ANOVA. Be sure to identify the independent and dependent variables.

3. Using SPSS, and using the data in Chapter 14 Data Set 2, complete the analysis and interpret the results for level of severity and type of treatment for pain relief. It is a 2 × 3 experiment, like the kind you saw in the answer to #2 above.

4. Use Chapter 14 Data Set 3 and answer the following questions in an analysis of whether level of stress and gender has an impact on caffeine consumption (measured as cups of coffee per day).

   a. Is there a difference between the levels of caffeine consumption for the high-stress, low-stress, and no-stress groups?

   b. Is there a difference between males and females, regardless of stress group?

   c. Any interactions?

15

Cousins or Just Good Friends?

Testing Relationships Using the Correlation Coefficient

Difficulty Scale ☺☺☺☺ (easy—
you don't even have to figure anything out!)

WHAT YOU'LL LEARN ABOUT IN THIS CHAPTER

✦ How to test the significance of the correlation coefficient

✦ The interpretation of the correlation coefficient

✦ The important distinction between significance and meaningfulness (again!)

✦ How to use SPSS to analyze correlational data and how to understand the results of the analysis

INTRODUCTION TO TESTING THE CORRELATION COEFFICIENT

In his research article on the relationship between the quality of a marriage and the quality of the relationship between the parent and the child, Daniel Shek tells us that there are at least two possibilities. First, a poor marriage might enhance parent-child relationships. This is because parents who are dissatisfied with their marriage might substitute their relationship with their children for emotional gratification. Or, according to the spillover hypothesis, a poor marriage might damage the parent-child relationship. This is because a poor marriage might set the stage for increased difficulty in parenting children.

Shek examined the link between marital quality and parent-child relationships in 378 Chinese married couples over a 2-year period. He found that higher levels of marital quality were related to higher levels of parent-child relationships; this was found for concurrent measures (at the present time) as well as longitudinal measures (over a period of time). He also found that the strength of the relationship between parents and children was the same for both mothers and fathers. This is an obvious example of how the use of the correlation coefficient gives us the information we need about whether sets of variables are related to one another. Shek computed a whole bunch of different correlations across mothers and fathers as well at Time 1 and Time 2, but all with the same purpose: to see if there was a significant correlation between the variables. Remember that this does not say anything about the causal nature of the relationship, only that the variables are associated with one another.

Want to know more? Check out Shek, D. T. L. (1998). Linkage between marital quality and parent-child relationship. *Journal of Family Issues, 19,* 687–704.

The Path to Wisdom and Knowledge

Here's how you can use the flow chart to select the appropriate test statistic, the test for the correlation coefficient. Follow along the highlighted sequence of steps in Figure 15.1.

1. The relationship between variables, and not the difference between groups, is being examined.

2. Only two variables are being used.

3. The appropriate test statistic to use is the *t* test for the correlation coefficient.

COMPUTING THE TEST STATISTIC

Here's something you'll probably be pleased to read: The correlation coefficient can act as its own test statistic. This makes things much

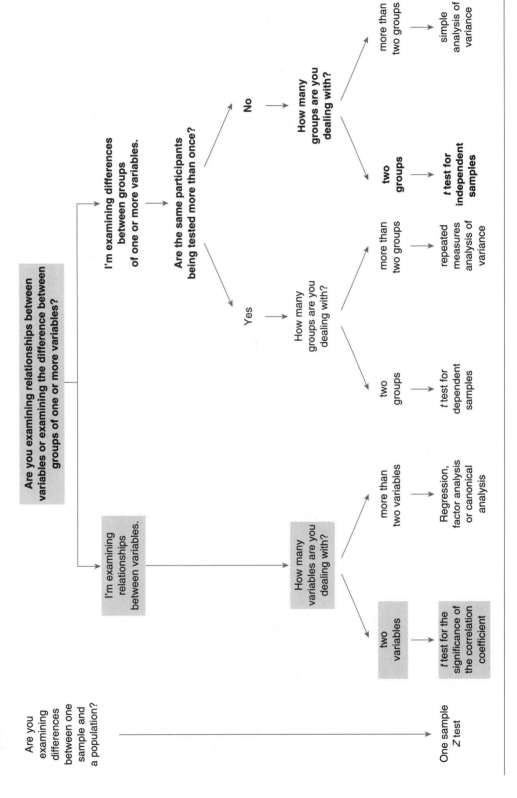

Figure 15.1 Determining That a *t* Test for the Correlation Coefficient Is the Correct Test Statistic

255

easier because you don't have to compute any test statistics, and examining the significance is very easy indeed.

Let's use, as an example, the following data that examine the relationship between two variables, the quality of marriage and the quality of parent-child relationships.

Quality of Marriage	Quality of the Parent-Child Relationship
76	43
81	33
78	23
76	34
76	31
78	51
76	56
78	43
98	44
88	45
76	32
66	33
44	28
67	39
65	31
59	38
87	21
77	27
79	43
85	46
68	41
76	41
77	48
98	56

Quality of Marriage	Quality of the Parent-Child Relationship
98	56
99	55
98	45
87	68
67	54
78	33

You can use Formula 5.1 from Chapter 5 (on page 81) to compute the Pearson correlation coefficient. When you do, you will find that $r = .393$. Now let's go through the steps of actually testing the value for significance and making a decision as to what the value means.

Here are the famous eight steps and the computation of the t-test statistic.

1. A statement of the null and research hypotheses.

The null hypothesis states that there is no relationship between the quality of the marriage and the quality of the relationship between parents and children. The research hypothesis is a two-tailed, nondirectional research hypothesis because it posits that there is a relationship between the two variables, but the direction is not important. Remember that correlations can be positive (or direct) or negative (or indirect), and the most important characteristic of a correlation coefficient is its absolute value or size and not its sign (positive or negative).

The null hypothesis is shown in Formula 15.1.

$$H_0: \rho_{xy} = 0 \qquad\qquad (15.1)$$

The Greek letter ρ, or rho, represents the population estimate of the correlation coefficient.

The research hypothesis (shown in Formula 15.2) states that there is a relationship between the two values, and that the relationship differs from a value of 0.

$$H_1: r_{xy} \neq 0 \qquad\qquad (15.2)$$

One tail or two? It's pretty easy to conceptualize what a one-tailed versus a two-tailed test is when it comes to differences between means. And it may even be easy for you to understand a two-tailed test of the correlation coefficient (where any difference from zero is what's tested). But what about a one-tailed test? It's really just as easy. A directional test of the research hypothesis that there is a relationship posits that relationship as being either direct (positive) or indirect (negative). So, if you think that there is a positive correlation between two variables, then the test is one-tailed. Similarly, if you hypothesize that there is a negative correlation between two variables, the test is one-tailed as well. It's only when you don't predict the direction of the relationship that the test is two-tailed. Got it?

2. Setting the level of risk (or the level of significance or Type I error) associated with the null hypothesis.

The level of risk or Type I error or level of significance is .05.

3. and 4. Selection of the appropriate test statistic.

Using the flow chart shown in Figure 15.1, we determined that the appropriate test is for the correlation coefficient. In this instance, we do not need to compute a test statistic because the sample r value (r_{xy} = .393) is, for our purposes, the test statistic.

5. Determination of the value needed for rejection of the null hypothesis using the appropriate table of critical values for the particular statistic.

Table B.4 lists the critical values for the correlation coefficient.

Our first task is to determine the degrees of freedom (df), which approximate the sample size. For this particular test statistic, the degrees of freedom are $n - 2$, or $29 - 2 = 27$, where n is equal to the number of pairs used to compute the correlation coefficient. These are the degrees of freedom only for this test statistic and not necessarily for any other.

Using this number (27), the level of risk you are willing to take (.05), and a two-tailed test (because there is no direction to the research hypothesis), the critical value is .381 (using $df = 25$ because it's more conservative than using 30). So, at the .05 level, with 27 degrees of freedom for a two-tailed test, the value needed for rejection of the null hypothesis is .381.

OK, we cheated a little. Actually, you can compute a *t* value (just like for the test for the difference between means) for the significance of the correlation coefficient. The formula is not any more difficult than any you have dealt with up to now, but you won't see it here. The point is that some smart statisticians have computed the critical *r* value for different sample sizes (and, likewise, degrees of freedom) for one- and two-tailed tests at different levels of risk (.01, .05), as you see in Table B.4. So, if you are reading along in your journal and see that a correlation was tested using a *t* value, you'll now know why.

6. A comparison of the obtained value and the critical value is made.

The obtained value is .393, and the critical value for rejection of the null hypothesis that the two variables are not related is .381.

7. and 8. Making a decision.

Now comes our decision. If the obtained value (or the value of the test statistic) is more extreme than the critical value (or the tabled value), the null hypothesis cannot be accepted. If the obtained value does not exceed the critical value, the null hypothesis is the most attractive explanation.

In this case, the obtained value (.393) does exceed the critical value (.381)—it is extreme enough for us to say that the relationship between the two variables (quality of marriage and quality of parent-child relationships) did occur by something other than chance.

So How Do I Interpret $r_{(27)} = .393, p < .05$?

- *r* represents the test statistic that was used
- 27 is the number of degrees of freedom
- .393 is the obtained value using the formula we showed you in Chapter 5
- $p < .05$ (the really important part of this little phrase) indicates that the probability is less than 5% on any one test of the null hypothesis that the relationship between the two variables is due to chance alone. Because we defined .05 as our criterion for the research hypothesis being more attractive than the null hypothesis, our conclusion is that there is a significant relationship between the two variables. This means that as the level of marital quality increases, so does the level of quality of the parent-child relationship. Similarly, as the level of marital quality decreases, so does the level of quality of the parent-child relationship.

> Correlation coefficients are used for lots of different purposes, and you're likely to read about them in journal articles being used to estimate the reliability of a test. But you already know all this because you read Chapter 6 and mastered it! In Chapter 6, you may remember that we talked about such reliability coefficients as test-retest (the correlation of scores at two points in time), parallel forms (the correlation between scores on different forms), and internal consistency (the intercorrelation between items). Correlation coefficients also are the standard units used in more advanced statistical techniques, such as those we discuss in Chapter 18.

Causes and Associations (Again!)

You'd have thought that you heard enough of this already, but this is so important that we really can't emphasize it enough. So, we'll emphasize it again. Just because two variables are related to one another (as in the previous example), it has no bearing on whether one causes the other. In other words, having a terrific marriage of the highest quality in no way ensures that the parent-child relationship will be of a high quality as well. These two variables may be correlated because they share some traits that might make a person a good husband or wife and also a good parent (patience, understanding, willingness to sacrifice), but it's certainly possible to see how someone can be a good husband or wife and have a terrible relationship with his or her children.

Remember the crimes and ice cream example from Chapter 5? It's the same here. Just because things are related and share something in common with one another has no bearing on whether there is a causal relationship between the two.

Significance Versus Meaningfulness (Again, Again!)

In Chapter 5, we reviewed the importance of the use of the coefficient of determination for understanding the meaningfulness of the correlation coefficient. You may remember that you square the correlation coefficient to determine the amount of variance accounted for by one variable in another variable. In Chapter 9, we also went over the general issue of significance versus meaningfulness.

But we should mention and discuss this topic again. Even if a correlation coefficient is significant (as was the case in the example in this chapter), it does not mean that the amount of variance accounted for is meaningful. For example, in this case, the coefficient of determination for a simple Pearson correlation value of

.393 is equal to .154, indicating that 15.4% of the variance is accounted for and a whopping 84.6% of the variance is not. It leaves lots of room for doubt, doesn't it?

So, even though we know that there is a positive relationship between the quality of a marriage and the quality of a parent-child relationship and they tend to "go" together, the relatively small correlation of .393 indicates that there are lots of other things going on in that relationship that may be important as well. So, if ever you wanted to apply a popular saying to statistics, "What you see is not always what you get."

USING THE COMPUTER TO COMPUTE A CORRELATION COEFFICIENT (AGAIN)

Here, we are using the data set named Chapter 15 Data Set 1, which has two measures of quality—one of marriage (time spent together in one of three categories) and one of parent-child relationships (strength of affection).

1. Enter the data in the Data Editor (or just open the file). Be sure you have two columns, each for a different variable. In Figure 15.2, you can see that the columns were labeled Qual_Marriage (for Quality of Marriage) and Qual_PC (Quality of Parent-Child Relationship).

Figure 15.2 Chapter 15 Data Set 1

	Qual_Marriage	Qual_PC
1	1	58.7
2	1	55.3
3	1	61.8
4	1	49.5
5	1	64.5
6	1	61.0
7	1	65.7
8	1	51.4
9	1	53.6
10	1	59.0

2. Click Analyze → Correlate → Bivariate, and you will see the Bivariate Correlations dialog box, as shown in Figure 15.3.

Figure 15.3   Bivariate Correlations Dialog Box

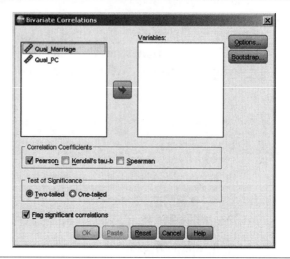

3. Double-click on the variable named Qual_PC to move it to the Variable(s) box, and then double-click on the variable named Qual_Marriage to move it to the Variable(s) box.

4. Click Two-tailed for a two-tailed test.

5. Click OK. The SPSS output is shown in Figure 15.4.

Figure 15.4   SPSS Output for Testing the Significance of the Correlation Coefficient

Correlations

Correlations

		Qual_Marriage	Qual_PC
Qual_Marriage	Pearson Correlation	1	.081
	Sig. (2-tailed)		.637
	N	36	36
Qual_PC	Pearson Correlation	.081	1
	Sig. (2-tailed)	.637	
	N	36	36

What the SPSS Output Means

This SPSS output is simple and straightforward.

The correlation between the two variables of interest is .081, which is significant at the .637 level, but to be (much) more precise, the probability of committing a Type I error is .637. That means that the likelihood of rejecting the null when true (that the two variables are not related) is about 63.7%—spookily high and very bad odds!

SUMMARY

Correlations are powerful tools that point out the direction of a relationship and help us to better understand what two different outcomes share with one another. Remember that correlations work only when you are talking about associations and never when you are talking about causal effects.

TIME TO PRACTICE

1. Given the following information, use Table B.4 in Appendix B to determine whether the correlations are significant and how you would interpret the results.

   a. The correlation between speed and strength for 20 women is .567. Test these results at the .01 level using a one-tailed test.

   b. The correlation between the number correct on a math test and the time it takes to complete the test is −.45. Test whether this correlation is significant for 80 children at the .05 level of significance. Choose either a one- or a two-tailed test and justify your choice.

   c. The correlation between number of friends and grade point average (GPA) for 50 adolescents is .37. Is this significant at the .05 level for a two-tailed test?

2. Use the data in Chapter 15 Data Set 2 to answer the questions below. Do the analysis manually or using SPSS.

   a. Compute the correlation between motivation and GPA.

   b. Test for the significance of the correlation coefficient at the .05 level using a two-tailed test.

   c. True or false? The more highly you are motivated, the more you will study. Which did you select and why?

3. Use the data in Chapter 15 Data Set 3 to answer the questions below. Do this one manually or use SPSS.

   a. Compute the correlation between income and level of education.

   b. Test for the significance of the correlation.

   c. What argument can you make to support the conclusion that "lower levels of education cause low income"?

4. Use the data in Chapter 15 Data Set 4 to answer the questions below. Compute the correlation coefficient manually.

   a. Test for the significance of the correlation coefficient at the .05 level using a two-tailed test between hours of studying and grade.

   b. Interpret this correlation. What do you conclude about the relationship between the number of hours spent studying and the grade received on a test?

   c. How much variance is shared between the two variables?

   d. How do you interpret the results?

5. A study was completed that examined the relationship between coffee consumption and level of stress for a group of 50 undergraduates. The correlation was .373, and it was a two-tailed test, done at the .01 level of significance. First, is the correlation significant? Second, what's wrong with the following statement? "As a result of the data collected in this study and our rigorous analyses, we have concluded that if you drink less coffee, you will experience less stress."

6. Use the following set of data to answer the questions. Do this one manually.

Age in Months	Number of Words Known
12	6
15	8
9	4
7	5
18	14
24	18
15	7
16	6
21	12
15	17

a. Compute the correlation between age in months and number of words known.

b. Test for the significance of the correlation at the .05 level of significance.

c. Go way back and recall what you learned in Chapter 5 about correlation coefficients and interpret this correlation.

7. Discuss the general idea that just because two things are correlated, it does not mean that one causes the other. Provide an example (other than ice cream and crime!).

16 Predicting Who'll Win the Super Bowl

Using Linear Regression

Difficulty Scale ☺ (as hard as they get!)

WHAT YOU'LL LEARN ABOUT IN THIS CHAPTER

✦ How prediction works and how it can be used in the social and behavioral sciences

✦ How and why linear regression works when predicting one variable from another

✦ How to judge the accuracy of predictions

✦ The usefulness of multiple regression

WHAT IS PREDICTION ALL ABOUT?

Here's the scoop. Not only can you compute the degree to which two variables are related to one another (by computing a correlation coefficient as we did in Chapter 5), but you can also use these correlations as the basis for the prediction of the value of one variable from the value of another. This is a very special case of how correlations can be used, and it is a very powerful tool for social and behavioral sciences researchers.

The basic idea is to use a set of previously collected data (such as data on variables X and Y), calculate how correlated these variables are with one another, and then use that correlation and the knowledge of X to predict Y. Sound difficult? It's not really, especially once you see it illustrated.

For example, a researcher collects data on total high school grade point average (GPA) and first-year college GPA for 400 students in their freshman year at the state university. He computes the correlation between the two variables. Then, he uses the techniques you'll learn about later in this chapter to take a *new* set of high school GPAs and (knowing the relationship between high school GPA and first-year college GPA from the previous set of students) predict what first-year GPA should be for a new sample of 400 students. Pretty nifty, huh?

Here's another example. A group of teachers is interested in finding out how well retention works. That is, do children who are retained in kindergarten (and not passed on to first grade) do better in first grade? Once again, these teachers know the correlation between being retained and first-grade performance; they can apply it to a new set of students and predict first-grade performance based on kindergarten performance.

How does regression work? Easy. Data are collected on past events (such as the existing relationship between two variables) and then applied to a future event given knowledge of only one variable. It's easier than you think.

The higher the absolute value of the correlation coefficient, the more accurate the prediction is of one variable from the other based on that correlation, because the more two variables share in common, the more you know about the second variable from your knowledge of the first variable. And you may already surmise that when the correlation is perfect (+1.0 or −1.0), then the prediction is perfect as well. If $r_{xy} = -1.0$ or $+1.0$, and if you know the value of X, then you also know the exact value of Y. Likewise, if $r_{xy} = -1.0$ or $+1.0$, and you know the value of Y, then you also know the exact value of X. Either way works just fine.

What we'll do in this chapter is go through the process of using linear regression to predict a Y score from an X score. We'll begin by discussing the general logic that underlies prediction, then go to a review of some simple line-drawing skills, and, finally, discuss the prediction process using specific examples.

THE LOGIC OF PREDICTION

Before we begin with the actual calculations and show you how correlations are used for prediction, let's create the argument why and how prediction works. Then, we will continue with the example of predicting college GPA from high school GPA.

Prediction is an activity that computes future outcomes from present ones. When we want to predict one variable from another, we need to first compute the correlation between the two variables. Table 16.1 shows the data we will be using in this example. Figure 16.1 shows the scatterplot (see Chapter 5) of the two variables that are being computed.

| Table 16.1 | Total High School GPA and First-Year College GPA Are Correlated | |

High School GPA	First-Year College GPA
3.50	3.30
2.50	2.20
4.00	3.50
3.80	2.70
2.80	3.50
1.90	2.00
3.20	3.10
3.70	3.40
2.70	1.90
3.30	3.70

To predict college GPA from high school GPA, we have to create a **regression equation** and use that to plot what is called a **regression line**. A regression line reflects our best guess as to what score on the Y variable (college GPA) would be predicted by a score on the X variable (high school GPA). For all the data you see in Table 16.1, it's the line that minimizes the distance between the line and each of the points on the predicted (Y) variable. You'll learn shortly how to draw that line shown in Figure 16.2. What does the regression line you see in Figure 16.2 represent?

Figure 16.1   Scatterplot of High School GPA and College GPA

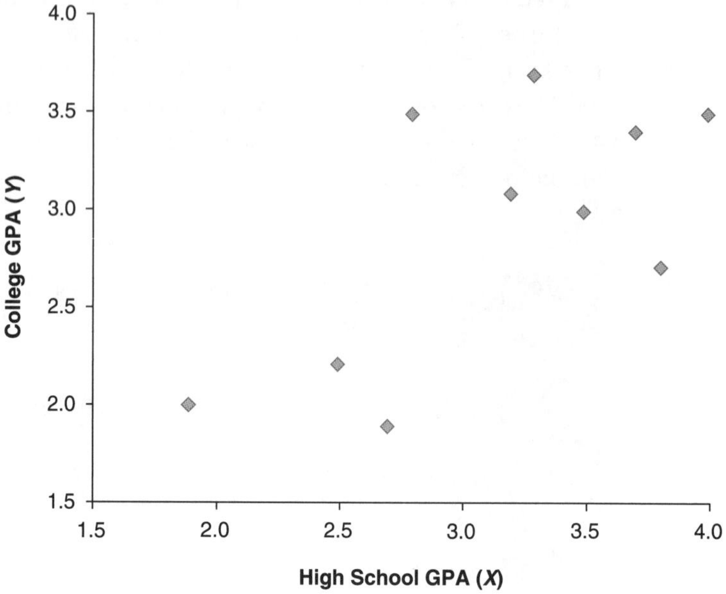

Figure 16.2   Regression Line of College GPA (Y) on High School GPA (X)

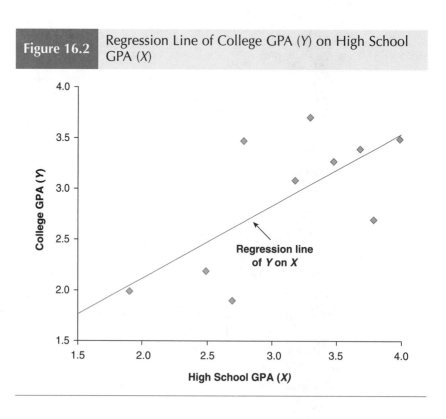

First, it's the regression of the *Y* variable on the *X* variable. In other words, *Y* (college GPA) is being predicted from *X* (high school GPA). This regression line is also called the **line of best fit**. The line best fits these data because it minimizes the distance between each individual point and the regression line. For example, if you take all of these points and try to find the line that best fits them all at once, the line you see in Figure 16.2 is the one you would use.

Second, it's the line that allows us our best guess (at estimating what college GPA would be, given each high school GPA). For example, if high school GPA is 3.0, then college GPA should be around (remember, this is only an eyeball prediction) 2.8. Take a look at Figure 16.3 to see how we did this. We located the predictor value (3.0) on the *x*-axis, then drew a perpendicular line from the *x*-axis to the regression line, then drew a horizontal line to the *y*-axis and *estimated* what the value would be.

Figure 16.3 Estimating College GPA Given High School GPA

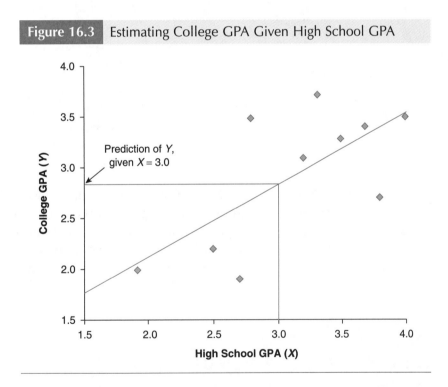

Third, the distance between each individual data point and the regression line is the **error in prediction**—a direct reflection of the correlation between the two variables. For example, if you look at data point 3.3, 3.7 (marked in Figure 16.4), you can see that this *X*, *Y* data point is above the regression line. The distance between that point and the line is the error in prediction, as marked in Figure 16.4, because if the prediction were perfect, then all the predicted points would fall where? Right on the regression or prediction line.

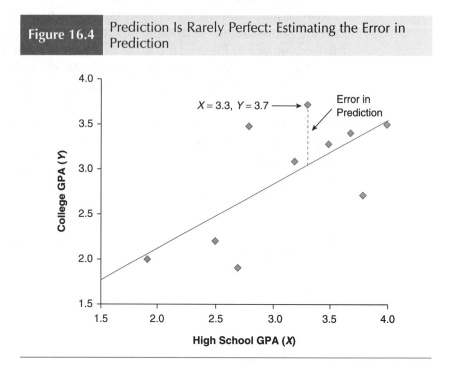

| Figure 16.4 | Prediction Is Rarely Perfect: Estimating the Error in Prediction |

Fourth, if the correlation were perfect, all the data points would align themselves along a 45° angle, and the regression line would pass through each point (just like we said in the third point above).

Given the regression line, we can use it to precisely predict any future score. That's what we'll do right now—create the line and then do some prediction work.

DRAWING THE WORLD'S BEST LINE (FOR YOUR DATA)

The simplest way to think of prediction is determining the score on one variable (which we'll call Y—the **criterion** or **dependent variable**) from the value of another score (which we'll call X—the **predictor** or **independent variable**).

The way that we find out how well X can predict Y is through the creation of the regression line we mentioned earlier in this chapter. This line is created from data that have already been collected. The equations are then used to predict scores using a new value for *X*, the predictor variable.

Formula 16.1 shows the general formula for the regression line, which may look familiar because you probably used it in your high school and college math courses. It's the same as the formula for any straight line.

$$Y' = bX + a \qquad (16.1)$$

where

Y' is the predicted score of *Y* based on a known value of *X*

b is the slope, or direction, of the line

a is the point at which the line crosses the *y*-axis

X is the score being used as the predictor

Let's use the same data shown earlier in Table 16.1 with a few more calculations that we will need thrown in.

X	*Y*	*X²*	*Y²*	*XY*
3.5	3.3	12.25	10.89	11.55
2.5	2.2	6.25	4.84	5.50
4.0	3.5	16.00	12.25	14.00
3.8	2.7	14.44	7.29	10.26
2.8	3.5	7.84	12.25	9.80
1.9	2.0	3.61	4.00	3.80
3.2	3.1	10.24	9.61	9.92
3.7	3.4	13.69	11.56	12.58
2.7	1.9	7.29	3.61	5.13
3.3	3.7	10.89	13.69	12.21
Total	31.4	29.3	102.50	89.99	94.75

ΣX or the sum of all the *X* values is 31.4

ΣY or the sum of all the *Y* values is 29.3

ΣX^2 or the sum of each *X* value squared is 102.5

ΣY^2 or the sum of each Y value squared is 89.99

ΣXY or the sum of the products of X and Y is 94.75

Formula 16.2 is used to compute the slope of the regression line (b in the equation for a straight line):

$$b = \frac{\Sigma XY - (\Sigma X \Sigma Y / n)}{\Sigma X^2 - [(\Sigma X)^2 / n]} \qquad (16.2)$$

In Formula 16.3, you can see the computed value for b, the slope of the line.

$$b = \frac{94.75 - [(31.4 \times 29.3)/10]}{102.5 - [(31.4)^2/10]}$$

$$b = \frac{2.749}{3.904} = .704 \qquad (16.3)$$

Formula 16.4 is used to compute the point at which the line crosses the y-axis (a in the equation for a straight line):

$$a = \frac{\Sigma Y - b\Sigma X}{n} \qquad (16.4)$$

In Formula 16.5, you can see the computed value for a, the intercept of the line.

$$a = \frac{29.3 - (.704 \times 31.4)}{10}$$

$$a = \frac{7.19}{10} = .719 \qquad (16.5)$$

Now, if we go back and substitute b and a into the equation for a straight line ($Y = bX + a$), we come up with the final regression line:

$$Y' = .704X + .719$$

Why the Y' and not just a plain Y? Remember, we are using X to predict Y and Y (read: **Y prime**) is the predicted and not the actual value of *Y*. So, now that we have this equation, what can we do with it? Predict *Y*, what else?

For example, let's say that high school GPA equals 2.8 (or *X* = 2.8). If we substitute the value of 2.8 into the equation, we get the following formula:

$$Y' = .704(2.8) + .719 = 2.69$$

So, 2.69 is the predicted value of *Y* (or *Y*) given *X* is equal to 2.8. Now, for any *X* score, we can easily and quickly compute a predicted *Y* score.

Tech Talk

Not all lines that fit best between a bunch of data points are straight. Rather, they could be curvilinear, just like you can have a curvilinear relationship, as we discussed in Chapter 5. For example, the relationship between anxiety and performance is such that when people are not at all anxious or very anxious, they don't perform very well. But if they're moderately anxious, then performance can be maximized. The relationship between these two variables is curvilinear, and the prediction of *Y* from *X* takes that into account.

How Good Is Our Prediction?

How can we measure how good a job we have done predicting one outcome from another? We know that the higher the absolute magnitude of the correlation between two variables, the better the prediction. In theory, that's great. But being practical, we can also look at the difference between the predicted value (*Y'*) and the actual value (*Y*) when we first compute the formula of the regression line.

For example, if the formula for the regression line is $Y' = .704X + .719$, the predicted *Y* (or *Y'*) for an *X* value of 2.8 is .704(2.8) + .719, or 2.69. We know that the actual *Y* value that corresponds to an *X* value is 3.5 (from the data set shown in Table 16.1). The difference between 3.5 and 2.69 is .81 and is known as an **error of estimate**.

If we take all of these differences, we can compute the average amount that each data point differs from the predicted data point, or the **standard error of estimate**. This value tells us how much imprecision there is in our estimate. As you might expect, the higher the correlation between the two values (and the better the prediction), the lower this error will be. In fact, if the correlation between the two variables is perfect (either +1 or −1), then the standard error of estimate is 0. Why? Because if prediction is perfect, all of the actual data points fall on the regression line, and there's no error in estimating *Y* from *X*.

> The predicted Y', or dependent variable, need not always be a continuous one, such as height, test score, or problem-solving skills. It can be a categorical variable, such as admit/don't admit, Level A/Level B, or Social Class 1/Social Class 2. The score that's used in the prediction is "dummy coded" to be a 1 or a 2 and then used in the same equation.

USING THE COMPUTER TO COMPUTE THE REGRESSION LINE

Let's use SPSS to compute the regression line in predicting Y from X. The data set we are using is Chapter 16 Data Set 1. We will be using the number of hours of training to predict how severe injuries will be if someone is injured playing football.

There are two variables in this data set:

Variable	Definition
Training (X)	Number of hours per week of strength training
Injuries (Y)	Severity of injuries on a scale from 1 to 10

Here are the steps to compute the regression line that we discussed in this chapter. Follow along and do it yourself.

1. Open the file named Chapter 16 Data Set 1.

2. Click Analyze → Regression → Linear. You'll see the Linear Regression dialog box shown in Figure 16.5.

Figure 16.5   Linear Regression Dialog Box

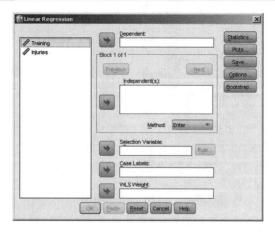

3. Click on the variable named Injuries, and click ⊡ to move it to the Dependent variable box. It's the dependent variable because its value depends on the value of number of hours of training. It's the variable being predicted.

4. Click on the variable named Training, and click ⊡ to move it to the Independent(s) variable box.

5. Click OK, and you will see the partial results of the analysis as shown in Figure 16.6.

Figure 16.6 Results of the SPSS Analysis

Coefficients[a]

Model		Unstandardized Coefficients		Standardized Coefficients	t	Sig.
		B	Std. Error	Beta		
1	(Constant)	6.847	1.004		6.818	.000
	Training	−.125	.046	−.458	−2.727	.011

a. Dependent Variable: Injuries

We'll get to the interpretation of this output in a moment. First, let's have SPSS overlay a regression line on the scatterplot for these data like the one you saw earlier in Figure 16.2.

6. Click Graphs → Legacy Dialogs → Scatter/Dot.

7. Click Simple Scatter, then click Define. You'll see the simple Scatterplot dialog box.

8. Click Injuries, and click ⊡ to move the variable label to the Y Axis box. Remember, the predicted variable is represented by the y-axis.

9. Click Training, and click ⊡ to move the variable label to the X Axis box.

10. Click OK, and you will see the scatterplot as shown in Figure 16.7.

Now let's draw the regression line.

1. If you are not into the chart editor, double-click on the chart to select it for editing.

Figure 16.7 A Scatterplot Generated Using SPSS

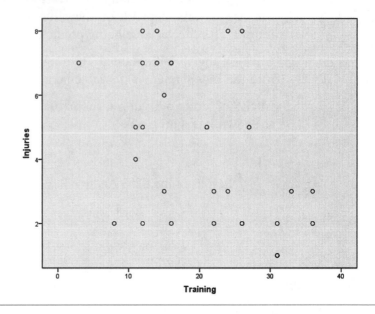

2. Click on the Add Fit Line at Total button that looks like this: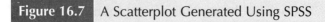

3. Click OK. The completed scatterplot, with the regression line, is shown in Figure 16.8 along with the multiple regression value R^2 equal to .21.

Figure 16.8 SPSS Scatterplot With Regression Line

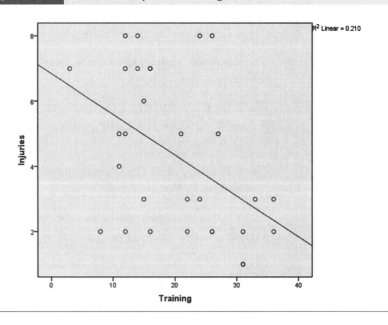

What the SPSS Output Means

The SPSS output tells us several things. First, the formula for the regression line is taken from the first set of output shown in Figure 16.6 as $Y' = -.125X + 6.847$. This equation can be used to predict level of injury given any number of hours spent in strength training. In fact, as you can see in Figure 16.8, the regression line has a negative slope, reflecting a negative correlation (of $-.458$) between hours of training and severity of injuries. So it appears, given the data, that the more one trains, the fewer severe injuries occur.

THE MORE PREDICTORS THE BETTER? MAYBE

All of the examples that we have used so far in the chapter have been for one criterion or outcome measure and one predictor variable. There is also the case of regression where more than one predictor or independent variable is used to predict a particular outcome. If one variable can predict an outcome with some degree of accuracy, then why couldn't two do a better job?

For example, if high school GPA is a pretty good indicator of college GPA, then how about high school GPA plus number of hours of extracurricular activities? So, instead of

$$Y' = bX + \alpha$$

the model for the regression equation becomes

$$Y' = bX_1 + bX_2 + \alpha$$

where

X_1 is the value of the first independent variable

X_2 is the value of the second independent variable

b is the regression weight for that particular variable

As you may have guessed, this model is called **multiple regression**. So, in theory anyway, you are predicting an outcome from two independent variables rather than one. But you want to add additional independent variables only under certain conditions.

First, any variable you add has to make a unique contribution to understanding the dependent variable. Otherwise, why use it? What do we mean by unique? The additional variable needs to explain differences in the predicted variable that the first predictor does not. That is, the two variables in combination would have to predict Y better than any one of the variables would do alone.

In our example, level of participation in extracurricular activities could make a unique contribution. But should we add a variable such as the number of hours each student studied in high school as a third independent variable or predictor? Because number of hours of study is probably highly related to high school GPA (another of our predictor variables, remember?), study time probably would not add very much to the overall prediction of college GPA. We might be better off looking for another variable (such as ratings on letters of recommendation) rather than spending our time collecting the data on study time.

The Big Rule(s) When It Comes to Using Multiple Predictor Variables

If you are going to use more than one predictor variable, try to keep the following two important guidelines in mind:

1. When selecting an independent variable to predict an outcome, select a predictor variable (X) that is related to the predicted variable (Y). That way, the two share something in common (remember, they should be correlated).

2. When selecting more than one independent variable (such as X_{IV1} and X_{IV2}), try to select variables that are independent or uncorrelated with one another, but are both related to the outcome or predicted (Y) variable.

In effect, you want only independent or predictor variables that are related to the dependent variable and are unrelated to each other. That way, each one makes as unique a contribution as possible in predicting the dependent or predicted variable.

How many predictor variables are too many? Well, if one variable predicts some outcome, and two is even more accurate, then why not three, four, or five predictor variables? In practical terms, every time you add a variable, an expense is incurred. Someone has to go collect the data, it takes time (which is $$$ when it comes to research budgets), and so on. From a theoretical sense, there is a fixed limit on how many variables can contribute to an understanding of what we are trying to predict. Remember that it is best when the predictor or independent variables are independent or unrelated to each other. The problem is that once you get to three or four variables, few things can remain unrelated. Better to be accurate and conservative than to include too many variables and waste money and the power of prediction.

SUMMARY

Prediction is a special case of simple correlations, and it is a very powerful tool for examining complex relationships. This might have been a little more difficult of a chapter than others, but you'll be well served by what you have learned, especially if you can apply it to the research reports and journal articles that you have to read. With the end of lots of chapters on inference, we're about to move on to using statistics when the sample size is very small or when the assumption that the scores are distributed in a normal way is violated.

TIME TO PRACTICE

1. How does linear regression differ from analysis of variance?

2. Chapter 16 Data Set 2 contains the data for a group of participants that took a timed test. The data are the average amount of time the participants took on each item (response time) and the number of guesses it took to get each item correct (number correct).

   a. What is the regression equation for predicting response time from number correct?

   b. What is the predicted response time if the number correct is 8?

   c. What is the difference between the predicted and the actual number correct for each of the predicted response times?

3. Betsy is interested in predicting how many 75-year-olds will develop Alzheimer's disease and is using as predictors level of education and general physical health graded on a scale from 1 to 10. But she is interested in using other predictor variables as well. Answer the following questions.

   a. What criteria should she use in the selection of other predictors? Why?

   b. Name two other predictors that you think might be related to the development of Alzheimer's disease.

c. With the four predictor variables (level of education and general physical health, and the two new ones that you name), draw out what the model of the regression equation would look like.

4. Go to the library and locate three different examples of where linear regression was used in a research study in your area of interest. It's OK if the study contains more than one predictor variable. Answer the following questions for each study.

a. What is one independent variable? What is the dependent variable?

b. If there is more than one independent variable, what argument does the researcher make that these variables are independent from one another?

c. Which of the three studies seems to present the least convincing evidence that the dependent variable is predicted by the independent variable, and why?

5. Here's where you can apply the information in one of this chapter's tips and get a chance to predict a Super Bowl winner! Joe Coach was curious to know if the average number of games won in a year predicts Super Bowl performance (win or lose). The X variable was the average number of games won during the past 10 seasons. The Y variable was whether the team ever won the Super Bowl during the past 10 seasons. Here are the data:

Team	Average Number of Wins Over 10 Years	Bowl? (1 = yes and 0 = no)
Savannah Sharks	12	1
Pittsburgh Pelicans	11	0
Williamstown Warriors	15	0
Bennington Bruisers	12	1
Atlanta Angels	13	1
Trenton Terrors	16	0
Virginia Vipers	15	1
Charleston Crooners	9	0
Harrisburg Heathens	8	0
Eaton Energizers	12	1

a. How would you assess the usefulness of the average number of wins as a predictor of whether a team ever won a Super Bowl?

b. What's the advantage of being able to use a categorical variable (such as 1 or 0) as a dependent variable?

c. What other variables might you use to predict the dependent variable, and why would you choose them?

6. Check your calculation of the correlation coefficient of the relationship between coffee consumption and stress done in Chapter 15, Question 5. If you wanted to know if coffee consumption predicts group membership . . .

   a. What is the predictor?

   b. What is the criterion?

   c. Do you have an idea as to what R^2 will be?

7. Time to try out multiple predictor variables. Take a look at the data shown here with the outcome being a great chef. We suspect that variables such as number of years of experience cooking, level of formal culinary education, and number of different positions (sous chef, pasta station, etc.) all contribute to rankings or scores on the Great Chef Test.

Years of Experience	Level of Education	# Positions	Score on Great Chef Test
5	1	5	88
6	2	4	78
12	3	9	56
21	3	8	88
7	2	5	97
9	1	8	90
13	2	8	79
16	2	9	85
21	2	9	60
11	1	4	89
15	2	7	88
15	3	7	76
1	3	3	78
17	2	6	98
26	2	8	91
11	2	6	88
18	3	7	90
31	3	12	98
27	2	16	88

By this time, you should be pretty much used to creating equations from data like these, so let's get to the real question.

   a. Which are the best predictors of the Chef's score?

   b. What can you expect for a score from a person with 12 years of experience and a Level 2 education, and who has held five positions?

17 What to Do When You're Not Normal

Chi-Square and Some Other Nonparametric Tests

Difficulty Scale ☺☺☺☺ (easy)

WHAT YOU'LL LEARN ABOUT IN THIS CHAPTER

✦ A brief survey of nonparametric statistics and when and how they should be used

INTRODUCTION TO NONPARAMETRIC STATISTICS

Almost every statistical test that we've covered so far in *Statistics for People Who (Think They) Hate Statistics* assumes that the data set with which you are working has certain characteristics. For example, one assumption underlying a *t* test between means (be the means independent or dependent) is that the variances of each group are homogeneous, or similar. And the assumptions can be tested. Another assumption of many **parametric statistics** is that the sample is large enough to represent the population. Statisticians have found that it takes a sample size of about 30 to fulfill this assumption. Many of the statistical tests we've covered so far are also robust, or powerful enough so that even if one of these assumptions is violated, the test is still valid.

But what do you do when the assumptions may be violated? The original research questions are certainly still worth asking and answering. That's when we use **nonparametric statistics** (also

285

called distribution-free statistics). These tests don't follow the same "rules" (meaning they don't require the same assumptions as the parametric tests we've reviewed), but the nonparametrics are just as valuable. The use of nonparametric tests also allows us to analyze data that come as frequencies, such as the number of children in different grades or the percentage of people receiving social security.

For example, if we wanted to know whether the number of people who voted for the school voucher in the most recent election is what we would expect by chance, or if there was really a pattern of preference, we would then use a nonparametric technique called *chi-square*.

In this chapter, we will cover chi-square, one of the most commonly used nonparametric tests, and provide a brief review of some others just so you can become familiar with some of the nonparametric tests that are available.

INTRODUCTION TO ONE-SAMPLE CHI-SQUARE

Chi-square is an interesting nonparametric test that allows you to determine if what you observe in a distribution of frequencies would be what you would expect to occur by chance. A one-sample chi-square includes only one dimension, such as the example you'll see here. A two-sample chi-square includes two dimensions, such as whether preference for the school voucher is independent of political party affiliation and gender.

For example, here are data from a sample selected at random from the 1990 census data collected in Sonoma County, California. As you can see, the table organizes information about level of education.

Level of Education			
No College	**Some College**	**College Degree**	**Total**
25	42	17	84

The question of interest here is whether the number of respondents is equally distributed across all levels of education. To answer this question, the chi-square value was computed and then tested for significance. In this example, the chi-square value is equal to 11.643, which is significant beyond the .05 level. The conclusion is that the number of respondents at the various levels of education for this sample is not equally distributed. In other words, it's not what we would expect by chance.

The rationale behind the one-sample chi-square test is that in any set of occurrences, you can easily compute what you would expect by chance. You do this by dividing the total number of occurrences by the number of classes or categories. In our census example above, the observed total number of occurrences was 84. We would expect that, by chance, 84/3 (84, which is the total of all frequencies, divided by 3, which is the total number of categories), or 28, respondents would fall into each of the three categories of level of education.

Then we look at how different what we expect by chance is from what we observe. If there is no difference between what we expect and what we observe, the chi-square value would be equal to zero.

Let's look more closely at how the chi-square value is computed.

COMPUTING THE CHI-SQUARE TEST STATISTIC

The chi-square test involves a comparison between what is observed and what would be expected by chance. The formula for computing the chi-square value for a one-sample chi-square test is shown in Formula 17.1.

$$\chi^2 = \sum \frac{(O-E)^2}{E} \tag{17.1}$$

where

χ^2 is the chi-square value

Σ is the summation sign

O is the observed frequency

E is the expected frequency

Here are some data we'll use to compute the chi-square value. Here are the famous eight steps to test this statistic.

Preference for School Voucher			
For	**Maybe**	**Against**	**Total**
23	17	50	90

1. A statement of the null and research hypotheses.

The null hypothesis shown in Formula 17.2 states that there is no difference in the frequency or the proportion of occurrences in each category.

$$H_0: P_1 = P_2 = P_3$$

(17.2)

The P in the null hypothesis represents the percentage of occurrences in any one category. This null hypothesis states that the percentages of cases in Category 1 (For), Category 2 (Maybe), and Category 3 (Against) are equal. We are using only three categories, but the number could be extended as the situation fits as long as each of the categories is mutually exclusive, meaning that any one observation cannot be in more than one category. For example, you can't be both male and female. Or, you can't be both For and Against the voucher plan at the same time.

The research hypothesis shown in Formula 17.3 states that there is a difference in the frequency or proportion of occurrences in each category.

$$H_1: P_1 \neq P_2 \neq P_3$$

(17.3)

2. Setting the level of risk (or the level of significance or Type I error) associated with the null hypothesis.

The Type I error rate is set at .05. How did we decide on this value and not some other, like .01 or 001? As we have emphasized in past chapters, we made the (somewhat) arbitrary decision to take this amount of risk.

3. Selection of the appropriate test statistic.

Any test between frequencies or proportions of mutually exclusive categories (such as For, Maybe, and Against) requires the use of chi-square. The flow chart we have used all along at the beginning of Chapters 10 through 16 to select the type of statistical test to use is not applicable to nonparametric procedures.

4. Computation of the test statistic value (called the obtained value).

Let's go back to our voucher data and construct a worksheet that will help us compute the chi-square value.

Category	O (observed frequency)	E (expected frequency)	D (difference)	$(O - E)^2$	$(O - E)^2/E$
For	23	30	7	49	1.63
Maybe	17	30	13	169	5.63
Against	50	30	20	400	13.33
Total	90	90			

Here are the steps we took to prepare this worksheet.

1. Enter the categories (Category) of For, Maybe, and Against. Remember that these three categories are mutually exclusive. Any data point can be in only one category at a time.

2. Enter the observed frequency (O), which reflects the data that were collected.

3. Enter the expected frequency (E), which is the total of the observed frequency (90) divided by the number of categories (3), or 90/3 = 30.

4. For each cell, subtract the expected frequency from the observed frequency (D). It does not matter which is subtracted from the other because these values are squared in the next step.

5. Square the observed minus the expected value. You can see these values in the column named $(O - E)^2$.

6. Divide the difference between the observed and the expected frequencies that have been squared, by the expected frequency. You can see these values in the column marked $(O - E)^2/E$.

7. Sum up this last column, and you have the total chi-square value of 20.6.

5. Determination of the value needed for rejection of the null hypothesis using the appropriate table of critical values for the particular statistic.

Here's where we go to Table B5 for the list of critical values for the chi-square test.

Our first task is to determine the degrees of freedom (df), which approximates the number of categories in which data have been organized. For this particular test statistic, the degrees of freedom are $r - 1$, where r equals rows, or $3 - 1 = 2$.

Using this number (2) and the level of risk you are willing to take (defined earlier as .05), you can use the chi-square table to look up the critical value. It is 5.99. So, at the .05 level, with 2 degrees of freedom, the value needed for rejection of the null hypothesis is 5.99.

6. A comparison of the obtained value and the critical value is made.

The obtained value is 20.6, and the critical value for rejection of the null hypothesis that the frequency of occurrences in Groups 1, 2, and 3 are equal is 5.99.

7. and 8. Decision time!

Now comes our decision. If the obtained value is more extreme than the critical value, the null hypothesis cannot be accepted. If the obtained value does not exceed the critical value, the null hypothesis is the most attractive explanation. In this case, the obtained value exceeds the critical value—it is extreme enough for us to say that the distribution of respondents across the three groups is not equal. Indeed, there is a difference in the frequency of people voting for, maybe, or against when it comes to preference for the school voucher.

Tech Talk

A commonly used name for the one-sample chi-square test is goodness of fit. This name suggests the question of how well a set of data "fits" an existing set. The set of data is, of course, what you observe. The "fit" part suggests that there is another set of data to which the observed set can be matched. This standard is the set of expected frequencies that are calculated in the course of computing the χ^2 value. If the observed data fit, it's just too close to what you would expect by chance and does not differ significantly. If the observed data do not fit, then what you observed is different from what you would expect.

So How Do I Interpret $\chi^2_{(2)} = 20.6$, $p < .05$?

- χ^2 represents the test statistic
- 2 is the number of degrees of freedom
- 20.6 is the obtained value using the formula we showed you earlier in the chapter
- $p < .05$ (the really important part of this little phrase) indicates that the probability is less than 5% on any one test of the null hypothesis that the frequency of votes is equally distributed across all categories by chance alone. Because we defined .05

as our criterion for the research hypothesis being more attractive than the null hypothesis, our conclusion is that there is a significant difference between the three sets of scores.

USING THE COMPUTER TO PERFORM A CHI-SQUARE TEST

Here's how to perform a simple, one-sample chi-square test using SPSS. We are using the data set named Chapter 17 Data Set 1, which was used in the school voucher example.

1. Open the data set. For a one-sample chi-square test, you need only enter the number of occurrences into each column, using a different value for each possible outcome. In this example, there would be a total of 90 data points in column 1: 23 would be entered as 1s (or For), 17 would be entered as 2s (or Maybe), and 50 would be entered as 3s (or Against).

2. Click Analyze → Nonparametric Tests → Legacy Dialogs → Chi-Square, and you will see the dialog box shown in Figure 17.1.

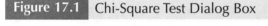

Figure 17.1 Chi-Square Test Dialog Box

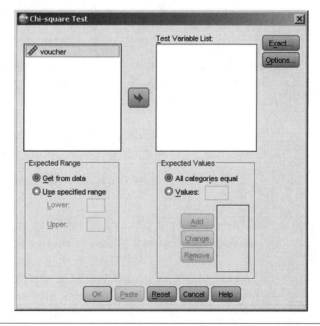

3. Double-click on the variable named voucher.

4. Click OK. SPSS will conduct the analysis and produce the output you see in Figure 17.2.

Figure 17.2 | SPSS Output for a Chi-Square Analysis

Chi-Square Test

Frequencies

Voucher

	Observed N	Expected N	Residual
1	23	30.0	−7.0
2	17	30.0	−13.0
3	50	30.0	20.0
Total	90		

Test Statistics

	Voucher
Chi-Square[a]	20.600
df	2
Asymp. Sig.	.000

a. 0 Cells (.0%) have expected frequencies less than 5. The minimum expected cell frequency is 30.0

What the SPSS Output Means

The SPSS output for the chi-square test shows you exactly what we talked about earlier. We'll address only the output that is relevant to our analysis.

1. The categories For (coded as 1), Maybe (coded as 2), and Against (coded as 3) are listed along with their Observed N.

2. This is followed by the Expected N, which in this case is 90/3, or 30.

3. The chi-square value of 20.600 and degrees of freedom appear under the Test Statistics section of the output.

The exact level of significance (in the figure, it's named Asymp. Sig.) is so small (less than .000) that SPSS computes it only as .000. A very unlikely outcome! So, it is highly unlikely that these three categories are equal in frequency.

OTHER NONPARAMETRIC TESTS YOU SHOULD KNOW ABOUT

You may never need a nonparametric test to answer any of the research questions that you propose. On the other hand, you may very well find yourself dealing with samples that may be very small (or at least fewer than 30) or data that violate some of the important assumptions underlying parametric tests.

Tech Talk

Actually, a primary reason why you may want to use nonparametric statistics is a function of the measurement level of the variable you are assessing. We'll talk more about that in the next chapter, but for now, most data that are categorical and are placed in categories (such as the Sharks and Jets) or that are ordinal and are ranked (1st, 2nd, and 3rd place) call for nonparametric tests of the kind you see in Table 17.1.

If that's the case, try nonparametrics on for size. Table 17.1 provides all you need to know about some other nonparametric tests, including their name, what they are used for, and a research question that illustrates how each might be used. Keep in mind that the table represents only a few of the many different tests that are available.

Table 17.1 Nonparametric Tests to Analyze Data in Categories and by Ranks

Test Name	When the Test Is Used	A Sample Research Question
To analyze data organized in categories		
McNemar test for significance of changes	To examine "before and after" changes	How effective is a phone call to undecided candidates on their voting for a particular issue?
Fisher's exact test	Computes the exact probability of outcomes in a 2 × 2 table	What is the exact likelihood of getting six heads on a toss of six coins?
Chi-square one-sample test (just like we focused on earlier in this chapter)	To determine if the number of occurrences across categories is random	Did brands Fruities, Whammies, and Zippies each sell an equal number of units during the recent sale?

(Continued)

Table 17.1	(Continued)

Test Name	When the Test Is Used	A Sample Research Question
To analyze data organized by ranks		
Kolmogorov-Smirnov test	To see whether scores from a sample came from a specified population	How representative is a set of judgments of other children of the entire elementary school to which they go?
The sign test, or median test	Used to compare the medians from two samples	Is the median income of people who voted for Candidate A greater than the median income of people who voted for Candidate B?
Mann-Whitney U test	Used to compare two independent samples	Did the transfer of learning, measured by number correct, occur faster for Group A than for Group B?
Wilcoxon rank test	To compare the magnitude as well as the direction of differences between two groups	Is preschool twice as effective as no preschool experience for helping develop children's language skills?
Kruskal-Wallis one-way analysis of variance	Compares the overall difference between two or more independent samples	How do rankings of supervisors differ between four regional offices?
Friedman two-way analysis of variance	Compares the overall difference between two or more independent samples on more than one dimension	How do rankings of supervisors differ as a function of regional office and gender?
Spearman rank correlation coefficient	Computes the correlation between ranks	What is the correlation between rank in the senior year of high school and rank during the freshman year of college?

SUMMARY

Chi-square is one of many different types of nonparametric statistics that help you answer questions based on data that violate the basic assumptions of the normal distribution or are just too small. These nonparametric tests are a very valuable tool, and even as limited an introduction as this will provide you with some assistance.

TIME TO PRACTICE

1. When does the obtained chi-square value equal 0? Provide an example of when this might happen.

2. Using the following data, test the question that an equal number of Democrats, Republicans, and Independents voted during the most recent election. Test the hypothesis at the .05 level of significance. Do this by hand.

Political Affiliation		
Republican	**Democrat**	**Independent**
800	700	900

3. Using the following data, test the question that an equal number of boys (code = 1) and girls (code = 2) participate in soccer at the elementary level at the .01 level of significance. (The data are available as Chapter 17 Data Set 2.) Use SPSS or some other statistical program and compute the exact probability of the chi-square value. What's your conclusion?

Gender	
Boys = 1	**Girls = 2**
45	55

4. School enrollment officials expected a change in the distribution of the number of students across grades and were not sure whether it is what they should have expected. Test the following data for goodness of fit at the .05 level. Here are the actual data . . .

Grade	1	2	3	4	5	6
Number of Students	309	432	346	432	369	329

18

Some Other (Important) Statistical Procedures You Should Know About

Difficulty Scale ☺☺☺☺ (moderately easy—just some reading and an extension of what you already know)

✦ An overview of more advanced statistical procedures and when and how they are used

Throughout *Statistics for People Who (Think They) Hate Statistics,* we covered only a small part of the whole body of statistics. We didn't have room, but more important, at the level at which you are beginning, it's important to keep things simple and direct.

However, that does not mean that in a research article you read or in some discussion in a class, you don't come across other analytical techniques that might be important for you to know about. So, for your edification, here are seven of those techniques, what they do, and examples of studies that used the technique to answer a question.

MULTIVARIATE ANALYSIS OF VARIANCE

You won't be surprised to learn that there are many different renditions of analysis of variance (ANOVA), each one designed to fit a

297

particular "the averages of more than two groups being compared" situation. One of these, multivariate analysis of variance (MANOVA), is used when there is more than one dependent variable. So, instead of looking just at one outcome, more than one outcome or dependent variable is used. If the dependent or outcome variables are related to one another (which they usually are—see the Tech Talk note in Chapter 13 about multiple *t* tests), it would be hard to determine clearly the effect of the treatment variable on any one outcome. Hence, MANOVA to the rescue.

For example, Jonathan Plucker from Indiana University examined gender, race, and grade differences in how gifted adolescents dealt with pressures at school. The MANOVA analysis that he used was a 2 (gender: male and female) × 4 (race: Caucasian, African American, Asian American, and Hispanic) × 5 (grade: 8th through 12th) MANOVA. The *multivariate* part of the analysis was the five subscales of the Adolescent Coping Scale. Using a multivariate technique, the effects of the independent variables (gender, race, and grade) can be estimated for each of the five scales, independent of one another.

Want to know more? Take a look at Plucker, J. A. (1998). Gender, race, and grade differences in gifted adolescents' coping strategies. *Journal for the Education of the Gifted, 21,* 423–436.

REPEATED MEASURES ANALYSIS OF VARIANCE

Here's another kind of analysis of variance. Repeated measures analysis of variance is very similar to any analysis of variance where, if you recall (see Chapter 13), the means of two or more groups are tested for differences. In a repeated measures ANOVA, there is one factor on which participants are tested more than once. That's why it's called repeated, because you repeat the process at more than one point in time on the same factor.

For example, B. Lundy, T. Field, C. McBride, T. Field, and S. Largie examined same-sex and opposite-sex interaction with best friends using juniors and seniors in high school. One of their main analyses was ANOVA with three factors: gender (male or female), friendship (same-sex or opposite-sex), and year in high school (junior or senior year). The repeated measure is year in high school, because the measurement was repeated across the same subjects.

Want to know more? Take a look at Lundy, B., Field, T., McBride, C., Field, T., & Largie, S. (1998). Same-sex and opposite-sex best friend interactions among high school juniors and seniors. *Adolescence, 33,* 130, 280–289.

ANALYSIS OF COVARIANCE

Here's our last rendition of ANOVA. Analysis of covariance (ANCOVA) is particularly interesting because it basically allows you to equalize initial differences between groups. Let's say you are sponsoring a program to increase running speed and want to compare how fast two groups of athletes can run a 100-yard dash. Because strength is often related to speed, you have to make some correction so that strength does not account for any differences at the end of the program. Rather, you want to see the effects of training with strength removed. You would measure participants' strength before you started the training program and then use ANCOVA to adjust final speed based on initial strength.

Michaela Hynie, John Lyndon, and Ali Tardash from McGill University used ANCOVA in their investigation of the influence of intimacy and commitment on the acceptability of premarital sex and contraceptive use. They used ANCOVA with social acceptability as the dependent variable (in which they were looking for group differences) and ratings of a particular scenario as the covariate. ANCOVA would ensure that differences in social acceptability would be corrected using ratings, so this would be one difference that would be controlled.

Want to know more? See Hynie, M., Lyndon, J., & Tardash, A. (1997). Commitment, intimacy, and women's perceptions of premarital sex and contraceptive readiness. *Psychology of Women Quarterly, 21,* 447–464.

MULTIPLE REGRESSION

You learned in Chapter 16 how the value of one variable can be used to predict the value of another. Often, social and behavioral sciences researchers look at how more than one variable can predict another. We touched on this in Chapters 5 and 16, and here's more about what is called multiple regression.

For example, it's fairly well established that parents' literacy behaviors (such as having books in the home) are related to how much and how well their children read. So, it would seem quite interesting to look at such variables as parents' age, education level, literacy activities, and shared reading with children to see what they contribute to early language skills and interest in books. Paula Lyytinen, Marja-Leena Laakso, and Anna-Maija Poikkeus did exactly that and used stepwise regression analysis to examine the contribution of parental background variables to children's literacy. They found that mothers' literacy activities and mothers' level of education contributed significantly to children's language skills, whereas mothers' age and shared reading did not.

Want to know more? Take a look at Lyytinen, P., Laakso, M. L., & Poikkeus, A. M. (1998). Parental contributions to child's early language and interest in books. *European Journal of Psychology of Education, 3,* 297–308.

FACTOR ANALYSIS

Factor analysis is a technique based on how well various items are related to one another and form clusters or factors. Each factor represents several different variables, and factors turn out to be more efficient than individual variables at representing outcomes in certain studies. In using this technique, the goal is to represent those things that are related to one another by a more general name, such as a factor. And the names you assign to these groups of variables called factors are not a willy-nilly process—the names reflect the content and the ideas underlying how they might be related.

For example, David Wolfe and his colleagues at the University of Western Ontario attempted to understand how experiences of maltreatment occurring before children were 12 years old affected peer and dating relationships during adolescence. To do this, the researchers collected data on many different variables and then looked at the relationship between all of them. Those that seemed to contain items that were related (and also belonged to a group that made theoretical sense) were deemed factors, such as the factor named Abuse/Blame in this study. Another factor was named Positive Communication and was made up of 10 different items, all of which were related to each other.

Want to know more? See Wolfe, D. A., Wekerle, C., Reitzel-Jaffe, D., & Lefebvre, L. (1968). Factors associated with abusive relationships among maltreated and non-maltreated youth. *Developmental Psychopathology, 10,* 61–85.

PATH ANALYSIS

Here's another statistical technique that examines correlations but allows a bit of a suggestion as to the direction, or causality, in the relationship between factors. Path analysis basically examines the direction of relationships through the postulation of some theoretical relationship between variables and then a test to see if the direction of these relationships is substantiated by the data.

For example, A. Efklides, M. Papadaki, G. Papantonious, and G. Kiosseoglou examined individual feelings of difficulty experienced in the learning of mathematics. To do this, they administered several different types of tests (such as those in the area of cognitive ability) and found that feelings of difficulty are mainly influenced by cognitive (problem-solving) rather than affective (emotional) factors. One of the most interesting uses of path analysis is that a technique called structural equation modeling is used to present the results in a graphical representation of the relationship between all of the different factors under consideration. That way, you can actually see what relates to what and with what degree of strength. Then, you can judge how well the data fit the model that was previously suggested. Cool.

Want to know more? Take a look at Efklides, A., Papadaki, M., Papantonious, G., & Kiosseoglou, G. (1998). Individual differences in feelings of difficulty: The case of school mathematics. *European Journal of Psychology of Education, 2,* 207–226.

STRUCTURAL EQUATION MODELING

Structural equation modeling (SEM) is a relatively new technique that has become increasingly popular since it was introduced in the early 1960s. Some researchers feel as if it is an umbrella term for techniques such as regression, factor analysis, and path analysis. Others believe that it stands on its own as an entirely separate approach. It's based on relationships between variables (like the previous three techniques we described).

The major difference between SEM and other advanced techniques such as factor analysis is that SEM is *confirmatory,* rather than *exploratory.* In other words, the researcher is more likely to use SEM to confirm whether a certain model that has been proposed works (meaning the data fit that model). Exploratory techniques set out to discover a particular relationship, with less (but not none) model building beforehand.

For example, Heather Gotham, Kenneth Sher, and Phillip Wood examined the relationships between young adult alcohol use disorders; preadulthood variables (gender, family history of alcoholism, childhood stressors, high school class rank, religious involvement, neuroticism, extraversion, psychoticism); and young adult developmental tasks (baccalaureate degree completion, full-time employment, marriage). Using structural equation modeling techniques, they found that preadulthood variables were more salient predictors of developmental tasks than a diagnosis of having a young adult alcohol use disorder.

Want to know more? Take a look at Gotham, H. J., Sher, K. J., & Wood, P. K. (2003). Alcohol involvement and developmental task completion during young adulthood. *Journal of Studies on Alcohol*, *64*(1), 32–42.

SUMMARY

Even though you probably will not be using these more advanced procedures anytime soon, that's all the more reason to know at least something about them, because you will certainly see them mentioned in various research publications and may even hear them mentioned in another class you are taking. And combined with your understanding of the basics (all the chapters in the book up to this one), you can really be confident of having mastered a good deal of important information about basic (and even some intermediate) statistics.

19 A Statistical Software Sampler

Difficulty Scale ☺☺☺☺☺ (a cinch!)

WHAT YOU'LL LEARN ABOUT IN THIS CHAPTER

✦ All about other types of software that allow you to analyze, chart, and better understand your data

---

You need not be a nerd or anything of the sort to appreciate and enjoy what the various computer programs can do for you in your efforts to learn and use basic statistics. The purpose of this chapter is to give you an overview of some of the more commonly used programs and some of their features, and a quick look at how they work. But before we go into these descriptions, here are some words of advice.

You can find a mega listing of software programs and links to the home pages of the companies that have created these programs at http://en.wikipedia.org/wiki/List_of_statistical_packages. We're only reviewing a few of the ones that we really like in this chapter, and if you are looking for a package (free or paid), take a look at this Wiki site and spend some time poking around. You can also find a very comprehensive list at http://www.statistics.com under Resources → Software.

SELECTING THE PERFECT STATISTICS SOFTWARE

Here are some tried-and-true suggestions for making sure that you get what you want from a stat program.

1. Whether the software program is expensive (like SPSS) or not (like EcStatic), be sure you try it out before you buy it. Almost every stat program listed offers a demo (usually on its website) that you can download, and in some cases, you can even ask them to send you a demo version on CD. These versions are often fully featured and last for up to 30 days, giving you plenty of time to try before you buy.

2. While we're mentioning price, buying directly from the manufacturer might be the *most* expensive way to go, especially if you buy outright without inquiring about discounts for students and faculty (what they sometimes call an educational discount). Your school bookstore may offer a discount, and a mail-order company might have even a better deal (again, ask about an educational discount). You can find these sellers' toll-free phone numbers listed in any popular computer magazine. You may also find that stat books (such as the SPSS version of this one) come with a limited/student version of the software, and in some cases, it is fully functioning and ready to go.

3. Many of the vendors who produce statistical analysis software offer two flavors. The first is the commercial version, and the second is the academic version. They are usually the same in content (but may be limited in very specific ways such as the number of variables you can test) and always differ, sometimes dramatically, in price. If you are going for the academic version, be sure that it is the same as the fully featured commercial version, and if not, then ask yourself if you can live with the differences. Why is the academic version so much cheaper? The company hopes that if you are a student, when you graduate, you'll move into some fat-cat job and buy the full version!

4. It's hard to know exactly what you'll need before you get started, but some packages come in modules, and you don't have to buy all of them to get the tools you need for the job you have to do. Read the company's brochures and website and call and ask questions.

5. Shareware is another option, and there are plenty of such programs available. Shareware is a method of distributing software so you pay for it only if you like it. Sounds like the honor system,

doesn't it? Well, it is. The prices are almost always very reasonable; the shareware is often better than the commercial product; and, if you do pay, you help ensure that the clever author will continue his or her other efforts at delivering new versions that are even better than the one you have.

6. Don't buy any software that does not offer telephone technical support or, at the least, some type of e-mail contact. To test this, call the tech support number (before you buy!) and see how long it takes for them to pick up the phone. If you're on hold for 20 minutes, that may indicate that they don't take tech support seriously enough to get to users' questions quickly. Or, if you email them and never hear back, look for another product.

7. Almost all the big stat packages do the same things—the difference is in the way that they do them. For example, SPSS, Minitab, and JMP all do a nice job of analyzing data and are acceptable. But it's the little things that might make a difference. For example, perhaps you want to import data from another software application. Some programs may allow this and others not. Be sure before you buy if any one specific feature is important to you.

8. Make sure you have the hardware to run the program you want to use. For example, most software is not limited by the number of cases and variables you want to analyze. The only limit is usually the size of your hard drive, which you'll use to store the data files. And if you have a slow machine (anything less than a Pentium) and less than 256 megabytes of RAM (random access memory), then you're likely to be waiting around and watching that hourglass while your CPU does its thing. Be sure of the hardware you need to run a program before you download the demo. Same goes for the version of your Mac operating system or your version of Windows—make sure all are compatible.

9. Operating systems are always changing, but sometimes software does not keep up. For example, some of these packages work only with a Windows operating system, and at that, some do not work with the newest version of Windows, Windows 7. If in doubt, call and be sure that your system has the stuff it takes to make these work. Some work with Windows and the Mac OS, and some now even have Linux versions.

10. Finally, some companies offer only downloads over the Internet, so you would not get an actual program disk. For the most part, this is fine, but for those of you who really, really, really like to cross your *t*s and dot your *i*s, it might make you a bit crazy not to have a DVD that you can call your own. Once again, check ahead.

WHAT'S OUT THERE

There are more statistical analysis programs available than you would ever need. Here's a listing of some of the most popular and their outstanding features. Remember that many of these do the same thing. If at all possible, as emphasized in the preceding section, try before you buy. Explore, explore, explore!

> The Wikipedia List of statistical packages (at http://en.wikipedia.org/wiki/List_of_statistical_packages) lists (and has links to many) more than 20 open source programs, five in the public domain, six Freeware, and more than 50 proprietary (which means you pay)—a great place to start in any hunt for what may be available and what may fit your specific or general needs.

First, the Free Stuff

Don't do anything until you've looked at the list of free statistical software available at the sites we mentioned above and at http://statpages.org/javasta2.html. Look under the "Completely Free" section. Also, don't ignore the "Free, but . . ." There are loads of packages here that you can download and use and that perform many of the procedures that you have learned about in *Statistics for People . . .* We can't possibly review all of them here, but spend some time tooling around and see what fits your needs.

My all-time favorite? OpenStat (at http://www.statpages.org/miller/openstat/), written by Dr. Bill Miller from Iowa State University. What's great about it? Well, first, it is entirely free—no "free for 124 days" stuff or anything like that. Next, it is very similar to SPSS and, in some ways, even easier to use (see Figure 19.1). It also features loads of options (far more than many commercial for-fee programs), including a bunch of nonparametric tests, measurement tools, and even tools for financial analysis (as in, how much can I save if I pay off my student loan early?). And finally, for those of you who like to tinker around, the "Open" in OpenStat means that it is open source. The program is written in C++, and if you know just a bit about that language, you can tweak the program as you see fit. It's a great all-around deal. And, on the site, you can learn about a new program that is compatible with Linux and the Mac OSX operating system.

| **Figure 19.1** | The OpenStat Data Entry Window |

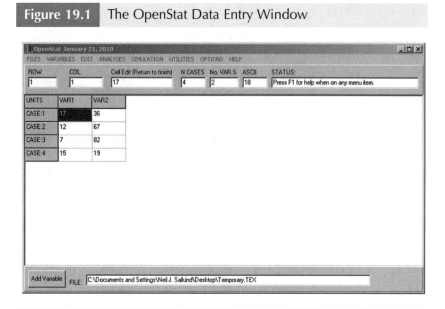

(Almost) SPSS

If you just can't shake that SPSS thing and really don't have hundreds of bucks to spare, take a look at PSPP (creative name, huh?) at http://www.gnu.org/software/pspp/, where there is an open source version of a program that mimics SPSS very closely. You can really do a lot here, at no charge, and get a very fair approximation of what SPSS looks like and does. And, if you like to tinker, the open source nature of this program (and this entire movement) welcomes you to actually change the code and make suggestions as to how to improve the program.

R

And if you are way ready for the big time computing, then turn to this open source program named R (probably after the two authors Robert Gentleman and Ross Ihaka). There are some commercial versions of R that offer formal support, but there is a huge community of R users that can be of help as well. And, R is also open source (like PSPP) and has its own journal (The R Journal); works on Linux, Unix, Windows, and Mac platforms; and takes some getting used to, but because it is command line controlled (which means you don't point and click but rather enter commands on a line), it is uniquely flexible in what you can tell it to do and how. Find everything you wanted to know about R, and download it as well, at http://www.r-project.org.

Time to Pay

JMP

JMP (now in Version 8) is billed as the "statistical discovery software." It operates on Mac, Windows, and Linux platforms and is software that "links statistics with graphics to interactively explore, understand, and visualize data." One of JMP's features is to present a graph accompanying every statistic so you can always see the results of the analysis both as statistical text and as a graphic. And all this is done automatically without you requesting it.

Need more information? Try http://www.jmp.com.

Cost: $1,595 for the single-user version and $595 for the academic version. The JMP Student Edition also appears in many textbooks. It is not available as an independent product.

Minitab

This is one of the first programs available for the personal computer, and it is now in Version 15 (it's been around a while!), which means that it's seen its share of changes over the years in response to users' needs. Some of the more outstanding features of this new version are

- Mentoring by Minitab with free initial consultation
- StatGuide™, which helps explain output
- ReportPad™, which is a report generator
- Online tutorials
- Easy-to-create graphs using One-click Graphs™

In Figure 19.2, you can see a sample of what Minitab output looks like for a simple bar chart and the main menus—neat and nicely organized. Folks, if you're going to spend $100, this is the one to look at carefully.

Need more information? Try http://www.minitab.com.

Cost: $99 for the full-fledged version, but there are various rental options for different prices for different time periods.

STATISTICA

StatSoft offers a family of STATISTICA products for Windows as well as STATISTICA for the Mac. Some of the features that are particularly nice about this powerful program are the self-prompting dialog boxes (you click OK, and STATISTICA tells you what to enter); the customization of the interface; easy integration with

Figure 19.2 | Minitab Output Showing a Simple Bar Chart

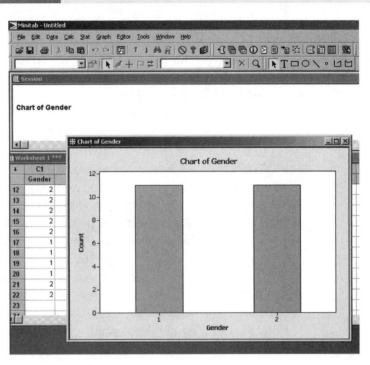

other programs; STATISTICA Visual Basic, which allows you to access more than 10,000 functions and use this development environment to design special applications; and the ability to use macros to automate tasks. A nice bonus at the website is an Electronic Statistics Textbook, which you can use to access information about many different topics.

Need more information? Try http://www.statsoftinc.com.

Cost: $795 for the base (and there are lots and lots of modules you can add) and $75 for the student version (which produces output with a watermark on it to remind you that you didn't spend $795).

SPSS

SPSS is one of the most popular big-time statistical packages in use, and that's why we have a whole appendix devoted to it in the very book you are holding in your hands! And also why we go through the computer exercises in these pages. It comes with a variety of different modules that cover all aspects of statistical analysis, including both basic and advanced statistics, and a version exists for almost every platform. It was recently purchased by IBM (after a brief renaming as PWAS, and thank heavens the new owners got rid of that).

Figure 19.3 shows some sample output for a simple descriptive and correlational analysis.

Need more information? Try http://www.spss.com.

Cost: Lots! $1,599 for the base version, $1,399 for the Mac version, and $599 for the academic base version of each.

SYSTAT

SYSTAT tends to be used by researchers in biological and physical sciences, whereas social and behavioral sciences researchers like SPSS (although the SYSTAT people are making an effort to appeal to the social and behavioral sciences people in the newest release, version 13, and even have a SPSS User? button that will help you make the transition). The really good news about SYSTAT is that its baby cousin is MYSTAT (at http://www.systat.com/MystatProducts.aspx), which is free and has lots of features—more than a beginner needs.

Need more information? Try http://www.systat.com.

Cost: $1,299 for the commercial version (sorry, Windows only!) and $499 for the academic version.

| **Figure 19.3** | Some Simple SPSS Analysis |

Correlations

Descriptive Statistics

	Mean	Std. Deviation	N
Training	20.10	8.833	30
Injuries	4.33	2.412	30

Correlations

		Training	Injuries
Training	Pearson Correlation	1	-.458[*]
	Sig. (2-tailed)		.011
	N	30	30
Injuries	Pearson Correlation	-.458[*]	1
	Sig. (2-tailed)	.011	
	N	30	30

*. Correlation is significant at the 0.05 level (2-tailed).

STATISTIX for Windows

Version 9 of STATISTIX offers a menu-driven interface that makes it particularly easy to learn to use and is almost as powerful as the other programs mentioned here. The company offers free technical support, and—ready for this?—a real, 450-page paper manual. And when you call technical support, you talk with the actual programmers, who know what they're talking about (my question was answered in 10 seconds!). Figure 19.4 shows you some STATISTIX output from a paired-sample *t* test. All around, a good deal.

Need more information? Try http://www.statistix.com.

Cost: $495 for the commercial version and $395 for the academic version, and both are Windows only.

EcStatic

The goal of the people at Someware Publishing in Vermont is to "provide intelligently crafted, easy-to-use statistical and graphing software at reasonable prices." They do this and more. EcStatic is a

Figure 19.4 STATISTIX Output for a Paired *t* Test

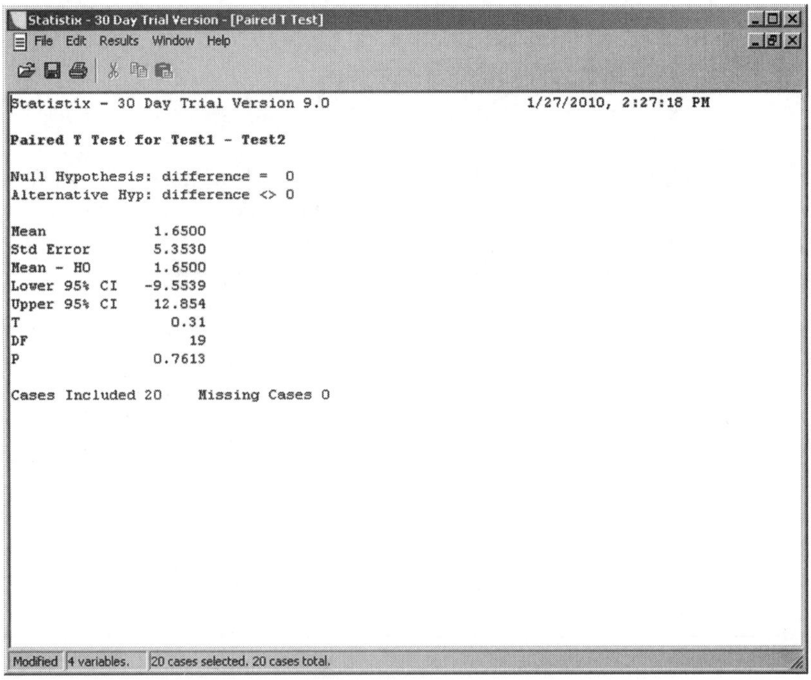

```
Statistix - 30 Day Trial Version - [Paired T Test]
File  Edit  Results  Window  Help

Statistix - 30 Day Trial Version 9.0                    1/27/2010, 2:27:18 PM

Paired T Test for Test1 - Test2

Null Hypothesis: difference =  0
Alternative Hyp: difference <> 0

Mean            1.6500
Std Error       5.3530
Mean - H0       1.6500
Lower 95% CI   -9.5539
Upper 95% CI    12.854
T                 0.31
DF                  19
P               0.7613

Cases Included 20    Missing Cases 0

Modified  4 variables.    20 cases selected. 20 cases total.
```

steal for the money. It is the least expensive of any of the programs that can really perform, and you certainly get much more than you pay for compared to the huge programs described previously. And if you think that this program is missing anything, take a look at the following list of some of the things it can do:

- Analysis of variance
- Breakdown
- Conversion of scores
- Correlation
- Cross-tabulation and chi-square
- Frequency distributions and histograms
- Nonparametric statistics
- Regression
- Scatterplot
- Summary statistics
- Transformations
- *t* test

Download a trial copy now and have some fun (and yes, there really is a link to Gus and Gertie's graphin' calculator on the site).

Need more information? Try http://www.somewareinvt.com.

Cost: $99.95 to download, with a discount on 10 or more ($49.95)—tell your instructor!

SUMMARY

That's the end of Part IV and just about the end of *Statistics for People Who (Think They) Hate Statistics*. But read on! The next chapter includes the ten best Internet sites in the universe for information about statistics, followed by Chapter 21, the ten commandments of data collection. Have fun with both of these.

PART V

Ten Things You'll Want to Know and Remember

Snapshots

"The Internet's got him ... Reboot! REBOOT!"

20

The Ten (or More) Best Internet Sites for Statistics Stuff

Way back in the first edition of *Statistics for People Who (Think They) Hate Statistics,* we told readers like you that if you're not yet using the Internet as a part of your learning and research activities, you are missing out on an extraordinary resource. Today, more than ever, students, researchers, and others certainly are taking advantage of this vast resource and feeling more comfortable doing it.

We all should recognize that what's on the Internet will not make up for a lack of studying or motivation—nothing will do that—but you can certainly find a great deal of information that will enhance your whole college experience. And this doesn't even begin to include all the fun you can have! But, as is well said, buyer beware. Although much of what you find on the Internet is valuable and accurate, certainly not all of this information has been verified. So, use and include at your own risk.

Now that you're a certified novice statistician, here are some Internet sites that you might find very useful should you want to learn more about statistics. Some are the same from last time, and some are new—have fun.

Although the locations of websites on the Internet are more stable than ever, they still can change frequently. The URL (uniform resource locator) that worked today might not work tomorrow. It's for this reason that you can find all of these websites (corrected and up-to-date) on the website for *Statistics for People Who (Think They) Hate Statistics* (at www.statisticsforpeople.com). Just go there and look for the resources link.

TONS AND TONS OF RESOURCES

You've been here before—Wikipedia (www.wikipedia.com). It's just about the best site for loads of information about anything, but happens to have particularly good reviews of important terms, examples, and more. A great community place to refer to and a great place to start as you begin a new topic. Many advanced terms and concepts are defined here, including examples. Just enter the search term and you're on your way.

For example, take a look at Figure 20.1, where you can see Wikipedia's entry for the arithmetic mean. Just like we did here and easy to jump to other links as well.

Figure 20.1	An Example Wikipedia Entry—Just One of 3,175,395 Articles (in English) and Growing

Arithmetic mean (AM)   [edit

Main article: Arithmetic mean

The *arithmetic mean* is the "standard" average, often simply called the "mean".

$$\bar{x} = \frac{1}{n} \cdot \sum_{i=1}^{n} x_i$$

The **mean** may often be confused with the median, mode or range. The mean is the arithmetic average of a set of values, or distribution; however, for skewed distributions, the mean is not necessarily the same as the middle value (median), or the most likely (mode). For example, mean income is skewed upwards by a small number of people with very large incomes, so that the majority have an income lower than the mean. By contrast, the median income is the level at which half the population is below and half is above. The mode income is the most likely income, and favors the larger number of people with lower incomes. The median or mode are often more intuitive measures of such data.

Nevertheless, many skewed distributions are best described by their mean – such as the exponential and Poisson distributions.

For example, the arithmetic mean of six values: 34, 27, 45, 55, 22, 34 is

$$\frac{34 + 27 + 45 + 55 + 22 + 34}{6} = \frac{217}{6} \approx 36.167.$$

WHO'S WHO AND WHAT'S HAPPENED?

The History of Statistics page, located at http://www.Anselm.edu/homepage/jpitocch/biostatshist.html, contains portraits and bibliographies of famous statisticians and a time line of important contributions to the field of statistics. So, names like Bernoulli, Galton, Fisher, and Spearman pique your curiosity? How about the development of the first test between two averages during the early 20th century? It might seem a bit boring until you have a chance to read about the people who make up this field and their ideas—in sum, pretty cool ideas and pretty cool people.

IT'S ALL HERE

SurfStat Australia (http://surfstat.anu.edu.au/surfstat-home/surfstat-main.html) is the online component of a basic stat course taught at the University of Newcastle, Australia, but has grown far beyond just the notes originally written by Annette Dobson in 1987, and updated over several years' use by Anne Young, Bob Gibberd, and others. Among other things, SurfStat contains a complete interactive statistics text. Besides the text, there are exercises, a list of other statistics sites on the Internet, and a collection of Java applets (cool little programs you can use to work with different statistical procedures).

HYPERSTAT

This online tutorial with 18 lessons, at http://www.davidmlane.com/hyperstat/index.html, offers nicely designed and user-friendly coverage of the important basic topics. What we really liked about the site was the glossary, which uses hypertext to connect different concepts to one another. For example, in Figure 20.2, you can see the definition of descriptive statistics also linked to other glossary terms, such as mean, standard deviation, and box plot. Click on any of those and zap! you're there.

Figure 20.2 Sample HyperStat Screen With Links to Other Terms

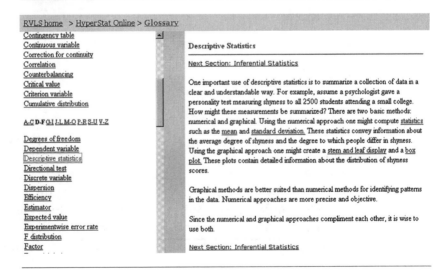

DATA? YOU WANT DATA?

There are data all over the place, ripe for the picking. Here are just a few. What to do with these? Download them to be used as examples in your work or as examples of analysis that you might want to do, and you can use these as a model.

- Statistical Reference Datasets at http://www.itl.nist.gov/div898/ strd/
- United States Census Bureau (a huge collection and a gold mine of data) at http://factfinder.census.gov/servlet/DatasetMain PageServlet?_lang=en
- The Data and Story Library, at http://lib.stat.cmu.edu/DASL/, with great annotations about the data (look for the stories link)
- Tons of economic data sets at Growth Data Sets (at http://www.bris.ac.uk/Depts/Economics/Growth/datasets.htm)

Then, there are all the data sets that are available through the federal government (besides the census). Your tax money supports them, so why not use them? For example, there's FEDSTATS (at http://www.fedstats.gov/), where more than 100 agencies in the U.S. federal government produce statistics of interest to the public. The Federal Interagency Council on Statistical Policy maintains this site to provide easy access to the full range of statistics and information produced by these agencies for public use. Here you can find country profiles contributed by the (boo!) CIA; public school student, staff, and faculty data (from the National Center for Education Statistics); and the Atlas of the United States Mortality (from the National Center for Health Statistics). What a ton of data!

FUN, REALLY FUN

So what do clever stat graduate students and their professors do for fun? Lots of things, but also they design these cute little applications (sometimes called Applets or Statlets) that let you jump in and start computing statistics with very little data entry or preparation. We really love the collection at http://www.bbn-school.org/us/math/ap_stats/applets/ applets.html, where you can click on Confidence Intervals and play with different values (such as sample size) to see the results, or click on one of the many links on this page to go to other lists of Applets.

MORE AND MORE AND MORE AND MORE RESOURCES

The University of Michigan's Statistical Resources on the Web (at http://www.lib.umich.edu/govdocs/stats.html) has hundreds and hundreds of resource links, including those to banking, book publishing, the elderly, and, for those of you with allergies, pollen count. Browse, search for what exactly it is that you need—no matter, you are guaranteed to find something interesting.

HOW ABOUT STUDYING STATISTICS IN STOCKHOLM?

The World Wide Web Virtual Library: Statistics is the name of the page, but the one-word title is misleading because the site (from the good people at the University of Florida at http://www.stat.ufl .edu/vlib/statistics.html) includes information on just about every facet of the topic, including data sources, job announcements, departments, divisions and schools of statistics (a huge description of programs all over the world), statistical research groups, institutes and associations, statistical services, statistical archives and resources, statistical software vendors and software, statistical journals, mailing list archives, and related fields. Tons of great information is available here. Make it a stop along the way.

ONLINE STATISTICAL TEACHING MATERIALS

If you do ever have to teach statistics, or even tutor fellow students, this is one place you'll want to visit: http://cedric.cnam.fr/PUBLIS/RC183.pdf for a very informative paper by Gilbert Saporta, including links to other resources. It contains valuable resources on every topic that was covered in *Statistics for People Who (Think They) Hate Statistics* and more. Then turn your attention to Teaching Statistics at http://www.math.grin.edu/~mooret/maa-notes/, where there are loads of links to everything from book reviews to tips and tricks.

MORE AND MORE AND MORE STUFF

Statistics.com (http://www.statistics.com) still has it all—a wealth of information on courses, tutoring, and even consulting. For example, if you wanted a listing of commercial (you pay) software that is available, take a look at Figure 20.3 to see just those listed under the letter S! Quite a few. You can then home in on the package that you might want to use (such as SPSS) and then learn everything from system requirements to performance to output options.

Figure 20.3	Just a Sample of the Information About Software Packages at statistics.com

AND FINALLY

We have the Internet Glossary of Statistics (at http://www.animated software.com/statglos/statglos.htm), where, if for some unknown reason you have not been able to use any of the above resources to find out an answer, you can come here and tap into definitions and tutorials.

21

The Ten Commandments of Data Collection

Now that you know how to analyze data, you would be well served to hear something about collecting them. The data collection process can be a long and rigorous one, even if it involves only a simple, one-page questionnaire given to a group of students, parents, patients, or voters. The data collection process may very well be the most time-consuming part of your project. But as many researchers do, this period of time is also used to think about the upcoming analysis and what it will entail.

Here they are: the ten commandments for making sure your data get collected in a way that they are usable. Unlike the original Ten Commandments, these should not be carved in stone (because they can certainly change), but if you follow them, you can avoid lots of aggravation.

Commandment 1. As you begin thinking about a research question, also begin thinking about the type of data you will have to collect to answer that question. Interview? Questionnaire? Paper and pencil? Find out how other people have done it in the past by reading the relevant journals in your area of interest and consider doing what they did. At least one of the lessons here is to not repeat others' mistakes. If something didn't work for them, it's most likely not going to work for you.

Commandment 2. As you think about the type of data you will be collecting, think about where you will be getting the data. If you are using the library for historical data or accessing files of data that have already been collected, such as census data (available through the U.S. Census Bureau and other locations online), you will have few logistical problems. But what if you want to assess the interaction between newborns and their parents? The attitude of teachers toward unionizing? The age at which people over 50 think they are old? All of these questions involve needing people to provide the answers, and finding people can be tough. Start now.

Commandment 3. Make sure that the data collection forms you use are clear and easy to use. Practice on a set of pilot data so you can make sure it is easy to go from the original scoring sheets to the data collection form. And, then have some colleagues complete the form to make sure it works.

Commandment 4. Always make a duplicate copy of the data file, and keep it in a separate location. Keep in mind that there are two types of people: those who have lost their data and those who will. Keep a copy of data collection sheets in a separate location. If you are recording your data as a computer file, such as a spreadsheet, be sure to make a backup! In fact, make two. These days, you can use online data backup services such as Carbonite (http://www.carbonite.com) or Mozy (http://www.mozy.com) in addition to your own physical backup.

Commandment 5. Do not rely on other people to collect or transfer your data unless you have personally trained them and are confident that they understand the data collection process as well as you do. It is great to have people help you, and it helps keep the morale up during those long data collection sessions. But unless your helpers are competent beyond question, you could easily sabotage all your hard work and planning.

Commandment 6. Plan a detailed schedule of when and where you will be collecting your data. If you need to visit three schools and each of 50 children needs to be tested for a total of 10 minutes at each school, that is 25 hours of testing. That does not mean you can allot 25 hours from your schedule for this activity. What about travel from one school to another? What about the child who is in the bathroom when it is his turn, and you have to wait 10 minutes until he comes back to the classroom? What about the day you show up and Cowboy Bob is the featured guest . . . and on and on. Be prepared for anything, and allocate 25%–50% more time in your schedule for unforeseen happenings.

Commandment 7. As soon as possible, cultivate possible sources for your subject pool. Because you already have some knowledge in your own discipline, you probably also know of people who work with the type of population you want or who might be able to help you gain access to these samples. If you are in a university community, it is likely that there are hundreds of other people competing for the same subject sample that you need. Instead of competing, why not try a more out-of-the-way (maybe 30 minutes away) school district or social group or civic organization or hospital, where you might be able to obtain a sample with less competition?

Commandment 8. Try to follow up on subjects who missed their testing session or interview. Call them back and try to reschedule. Once you get in the habit of skipping possible participants, it becomes too easy to cut the sample down to too small a size. And you can never tell—the people who drop out might be dropping out for reasons related to what you are studying. This can mean that your final sample of people is qualitatively different from the sample with which you started.

Commandment 9. Never discard the original data, such as the test booklets, interview notes, and so forth. Other researchers might want to use the same database, or you may have to return to the original materials for further information.

And number 10? Follow the previous 9. No kidding!

Snapshots

"Statistimagic takes old, haphazard data and turns it into sparkling new, *p*-is-less-than-five-percent *results*."

APPENDIX A

SPSS Statistics in Less Than 30 Minutes

This appendix will teach you enough about SPSS Statistics (let's just call it SPSS) to complete the exercises in *Statistics for People Who (Think They) Hate Statistics*. Learning SPSS is not rocket science—just take your time, work as slowly as you need to, and ask a fellow student or your instructor for help if necessary.

You are probably familiar with other Windows applications, and you will find that many SPSS features operate exactly the same. We assume you know about dragging, clicking, double-clicking, and working with Windows. If you do not, you can refer to one of the many popular trade computer books for help. SPSS works with Microsoft WindowsXP, Vista, or 7. It will not work with any earlier versions of Windows.

This appendix is an introduction to SPSS (Version 19.x) and shows you just some of the things it can do. Almost all of the information in this appendix also can be applied to earlier versions of SPSS, from version 11 through the current version, 19.

Throughout the examples in this appendix, we will use the sample data set shown in Appendix C named Sample Data. You are welcome to enter those data manually or download the set from the SAGE website at www.sagepub.com/salkind4e or the author's website at www.onlinefilefolder.com. The username is *ancillaries* and the password is *files*.

STARTING SPSS

Like other Windows-based applications, SPSS is organized as a group and is available on the Start menu. This group was created when you first installed SPSS. To start SPSS, follow these steps:

325

1. Click Start, then point to Programs.

2. Find and click the SPSS icon. When you do this, you will see the SPSS opening screen as shown in Figure A.1. You should note that some computers are set up differently, and your SPSS icon might be located on the desktop. In that case, to open SPSS, just double-click on the icon.

Figure A.1   The Opening SPSS Screen

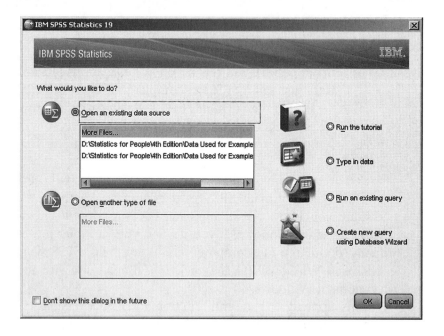

The SPSS Opening Window

As you can see in Figure A.1, the opening window presents a series of options that allows you to select from running the SPSS tutorial, entering data, posing an established query, creating a new query using the Database Wizard, or opening an existing source of data (an existing file). Should you not want to see this screen each time you open SPSS, then click on the Don't show this dialog in the future box in the lower left corner of the window.

For our purposes, we will click the Type in data option and click OK because it is likely to be the one you first select upon opening and learning SPSS. Once you do this, the Data View window (also called the Data Editor) you see in Figure A.2 becomes active. This is where you enter data you want to use with SPSS once those data have been defined. Although you cannot see it when SPSS first opens, there is another open (but not active) window as well. This is the Variable View, where variables are defined and the parameters for those variables are set.

Figure A.2	The Data View Window

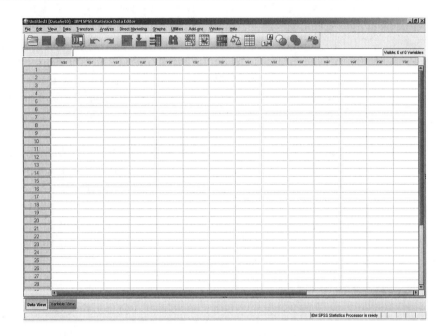

The Viewer displays statistical results and charts that you create. An example of the Viewer window is shown in Figure A.3. A data set is created using the Data Editor, and once the set is analyzed or graphed, you examine the results of the analysis in the Viewer.

Figure A.3	The Viewer

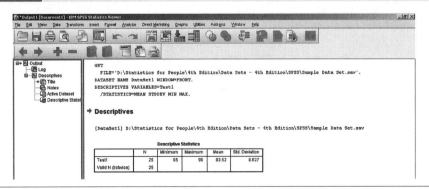

If you think the Data Editor is similar to a spreadsheet in form and function, you are right. In form, it certainly is, because the Data Editor consists of rows and columns just like in Excel. Values can be entered and then manipulated. In function as well, the Data Editor is much like a spreadsheet. Values that are entered can be transformed, sorted, rearranged, and more.

THE SPSS TOOLBAR AND STATUS BAR

The use of the Toolbar, the set of icons that is underneath the menus, can greatly facilitate your SPSS activities. If you want to know what an icon on the Toolbar does, just place the mouse pointer on it, and you will see a tip telling you what the tool does. Some of the buttons on the Toolbar are dimmed, meaning they are not active.

The Status Bar, located at the bottom of the SPSS window, is another useful on-screen tool. Here, you can see a one-line report regarding in which activity SPSS is currently involved. The message *SPSS for Windows processor is ready* tells you that SPSS is ready for your directions or input of data. Or, *Running Means . . .* tells you that SPSS is in the middle of the procedure named Means.

USING SPSS HELP

If you need help, you have come to the right place. SPSS offers help that is only a few mouse clicks away, and it is especially useful when you are in the middle of a data file and need information about an SPSS feature. SPSS Help is so comprehensive that even if you are a new SPSS user, it can show you the way.

You can get help in SPSS by pressing the F1 function key (see Figure A.2) or using the Help menu you see in Figure A.4.

| Figure A.4 | The Various Help Options |

There are 10 options on the Help menu, greatly expanded from earlier versions of SPSS, and 6 are directly relevant to helping you.

- Topics gives you a list of topics for which you can get help.
- Tutorial offers you a short tutorial on all aspects of using SPSS.
- Case Studies gives you real live examples of how SPSS can be applied.
- Statistics Coach walks you through procedures step by step.
- Command Syntax Reference helps you to learn and use SPSS's programming language.
- Developer Central provides you with information about creating add-ons and other program enhancements for SPSS.
- About . . . gives you some technical information about SPSS, including the current version on which you are working.
- Algorithms focuses on actual calculations that are used to produce the results you see in SPSS.
- SPSS Inc. Home takes you to the home page for SPSS on the Internet.
- Check for Updates automatically connects to the SPSS mother ship to check if there are any updates you should be aware of.

A BRIEF TOUR OF SPSS

Now, sit back and enjoy a brief tour of what SPSS can do. Nothing fancy here. Just some simple descriptions of data, a test of significance, and a graph or two. What we are trying to show you is how easy it is to use SPSS.

Opening a File

You can enter your own data to create a new SPSS data file, use an existing file, or even import data from such applications as Microsoft Excel into SPSS. Any way you do it, you need to have data to work with. In Figure A.5, the data contained in Appendix C are shown, called Sample Data, and they are also available on the Internet site.

Figure A.5	An Open SPSS File

Sample Data Set.sav [DataSet1] - IBM SPSS Statistics Data Editor

File   Edit   View   Data   Transform   Analyze   Direct Marketing   Graphs   Utilities   Add-

	ID	Gender	Treatment	Test1	Test2
1	1	Male	Control	98	32
2	2	Female	Experimental	87	33
3	3	Female	Control	89	54
4	4	Female	Control	88	44
5	5	Male	Experimental	76	64
6	6	Male	Control	68	54
7	7	Female	Control	78	44
8	8	Female	Experimental	98	32
9	9	Female	Experimental	93	64
10	10	Male	Experimental	76	37
11	11	Female	Control	75	43
12	12	Female	Control	65	56
13	13	Male	Control	76	78
14	14	Female	Control	78	99
15	15	Female	Control	89	87
16	16	Female	Experimental	81	56
17	17	Male	Control	78	78
18	18	Female	Control	83	56
19	19	Male	Control	88	67
20	20	Female	Control	90	88
21	21	Male	Control	93	81
22	22	Male	Experimental	89	93
23	23	Female	Experimental	86	87
24	24	Male	Control	77	80
25	25	Male	Control	89	99

A Simple Table and Graph

Now it is time to get to the reason why we are using SPSS in the first place—the various analytical tools that are available.

First, let's say we want to know the general distribution of males and females. That is all, just a count of how many males and how many females are in the total sample we are working with. We also want to create a simple bar graph of the distribution.

In Figure A.6, you will see the output that provides exactly the information we asked for, which was the frequency of the number of males and females. We used the Frequencies option on the Descriptive Statistics (under the main menu Analyze) to compute these values. Then, we used the Graphs option to create a simple bar graph of the frequency, as you can also see in Figure A.7.

| Figure A.6 | The Results of a Simple Descriptive Analysis |

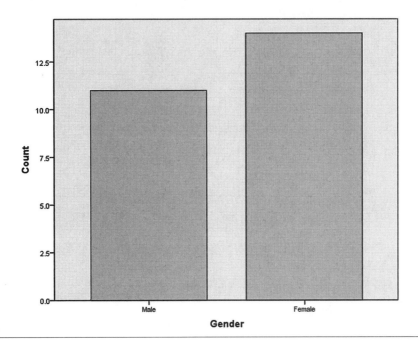

A Simple Analysis

Let's see if males and females differ in their average test1 scores. This is a simple analysis requiring a *t* test for independent samples. The procedure is a comparison between males and females for the mean of test1 for each group.

In Figure A.7, you can see a partial summary of the results of the *t* test. Notice that the listing in the left pane (the outline view) of the SPSS Viewer now shows the Frequencies, Graph, and *t*-test procedures listed. To see any part of the output, all we need do is click on that element. Almost always, when SPSS produces output in the Viewer, you will have to scroll to see the entire output.

| Figure A.7 | The Results of an Independent Samples *t* Test |

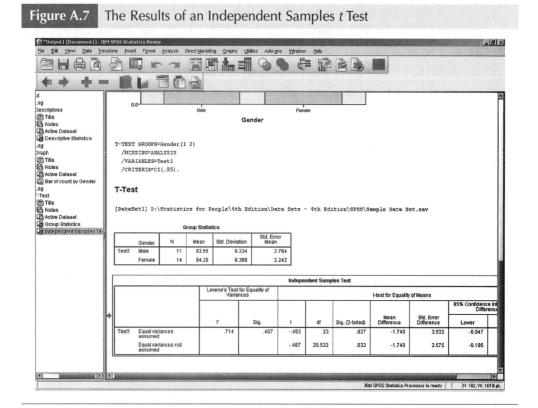

CREATING AND EDITING A DATA FILE

As a hands-on exercise, let's create the beginning of the sample data file you see in Appendix C. The first step is to define the variables in your data set and then to enter the data. You should have a new Data Editor window open (Click File → New → Data).

Defining Variables

SPSS cannot work unless variables are defined. You can have SPSS define the variables for you, or you can do the defining yourself, thereby having much more control over the way things look and work. SPSS will automatically name the first variable VAR00001. If you defined a variable in row 1, column 5, then SPSS would name the variable VAR00005 and also number the other columns sequentially. But you can also define variables, assigning a name of your choice.

Custom Defining Variables: Using the Variable View Window

In order to define a variable, you must first go to the Variable View window by clicking the Variable View tab at the bottom of the SPSS screen. Once that is done, you will see the Variable View window, as shown in Figure A.8, and be able to define any one variable as you see fit.

| Figure A.8 | The Variable View Window |

Once in the Variable View window, you can define variables along the following parameters:

- Name provides a name for a variable up to eight characters.
- Type defines the type of variable, such as text, numeric, string, scientific notation, and so on.
- Width defines the number of characters wide that the column housing the variable will occupy.
- Decimals defines the number of decimals that will appear in the Data View.
- Label defines a label up to 256 characters for the variable.
- Values defines the labels that correspond to certain numerical values (such as 1 for male and 2 for female).
- Missing indicates how missing data will be dealt with.
- Columns defines the number of spaces allocated for the variable in the Data View window.
- Align defines how the data are to appear in the cell (right, left, or center aligned).
- Measure defines the scale of measurement that best characterizes the variable (nominal, ordinal, or interval).
- Role defines the part that the variable plays in the overall analysis (input, target, etc.)

If you place the cursor in the first cell under the Name column, enter any name, and press the Enter key, then SPSS will automatically provide you with the default values for all the variable characteristics. Even if you are not in the Data View screen (click the tab on the bottom of the window), SPSS will automatically name the variables var0001, var0002, and so on.

In the Variable View, enter the names of the variables as you see in Figure A.9.

Figure A.9   Defining Variables in the Variable View Window

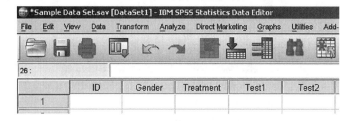

Now, if you wanted, you could switch to the Data View (see Figure A.10) and just enter the data as you see in Figure A.5. But first, let's look at just one of the cool SPSS bells and whistles.

Figure A.10   The Data View Window Ready for Data Entry

	ID	Gender	Treatment	Test1	Test2
1					

Defining Variable Labels

You can leave your data appearing as numerical values in the SPSS Data Editor, or you can have labels represent the numerical values (as you saw in Figure A.5).

Why would you want to change the label of a variable? You probably already know that, in general, it makes more sense to work with numbers (like 1 or 2) than with string or alphanumeric variables (such as male or female).

But it sure is a lot easier to look at a data file and see words rather than numbers. Just think about the difference between data files with numbers representing various levels (such as 1 and 2) of a variable and with the actual values (such as male and female). The Values option in the Variable View screen allows you to enter *values* in the cell, but what you will see are value *labels*.

If you click the ellipsis button in the Values column (see Figure A.11), you will see the Value dialog box, as shown in Figure A.12.

Figure A.11 The Values Column in the Variable View Screen

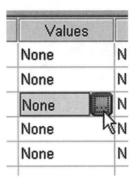

Figure A.12 The Value Labels Dialog Box

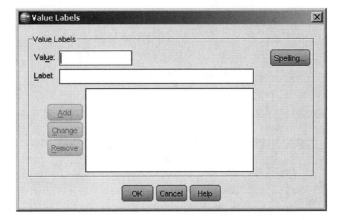

Changing Variable Labels

To assign or change a variable label, follow these steps. Here, we will label males as 1 and females as 2.

1. For the variable gender, click on the ellipsis (see Figure A.11) to open the Value Labels dialog box.

2. Enter a value for the variable, which, in this case, will be 1 for males.

3. Enter the value label for the value, which is male.

4. Click Add.

5. Do the same for female and value 2. When you finish your business in the Define Labels dialog box (see Figure A.13), click OK, and the new labels will take effect.

Figure A.13 The Completed Value Labels Dialog Box

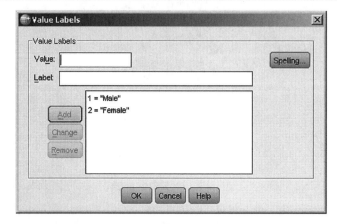

When you select View → Variable Labels from the main menu (in the Data View), you will see the labels in the Data Editor. Notice how the value of the entry in Figure A.14 is actually 2, even though the label in the cell reads Female.

Figure A.14 Seeing Variable Labels

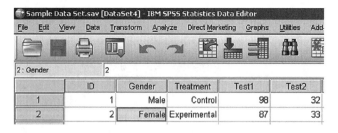

Opening a Data File

Once a file is saved, you have to open or retrieve it when you want to use it again. The steps are simple.

1. Click File → Open. You will see the Open Data File dialog box.

2. Find the data file you want to open, and highlight it.

3. Click OK.

A quick way to find and open an SPSS file is by clicking on its name at the bottom of the File menu. SPSS lists the most recently used files there.

PRINTING WITH SPSS

Here comes information on the last thing you will do once a data file is created. Once you have created the data file you want or completed any type of analysis or chart, you probably will want to print out a hard copy for safekeeping or for inclusion in a report or paper. Then, when your SPSS document is printed and you want to stop working, it is time to exit SPSS.

Printing is almost as important a process as editing and saving data files. If you cannot print, you have nothing to take away from your work session. You can export data from an SPSS file to another application, but getting a hard copy directly from SPSS is often more timely and more important.

Printing an SPSS Data File

It is simple to print either an entire data file or a selection from one.

1. Be sure that the data file you want to print is the active window.

2. Click File → Print. When you do this, you will see the Print dialog box.

3. Click OK, and whatever is active will print.

As you can see, you can choose to print the entire document or a specific selection (which you will have already made in the Data Editor window), and to increase the number of copies from 1 to 99 (the max number of copies you can print). You can also configure the print dialog box so that a PDF file is produced.

Printing a Selection From an SPSS Data File

Printing a selection from a data file follows exactly the steps that we listed above for printing a data file, except that in the Data Editor window, you select what you want to print and click on the Selection option in the Print dialog box. The steps go like this:

1. Be sure that the data you want to print are selected.
2. Click File → Print.
3. Click Selection in the Print dialog box.
4. Click OK, and whatever you selected will be printed.

CREATING AN SPSS CHART

A picture is worth a thousand words, and SPSS offers you just the features to create charts that bring the results of your analyses to life. In this part of Appendix A, we will go through the steps to create several different types of charts and provide examples of different charts. Then, we will show you how to modify a chart, including adding a chart title; adding labels to axes; modifying scales; and working with patterns, fonts, and more. For whatever reason, SPSS uses the words "graphs" and "charts" interchangeably.

Creating a Simple Chart

The one thing that all charts have in common is that they are based on data. Although you may import data to create a chart, in this example, we will use the data from Appendix C to create a bar chart (like the one you saw in Figure A.6) of the number of males and females in each group.

Creating a Bar Chart

The steps for creating any chart are basically the same. You first enter the data you want to use in the chart, select the type of chart you want from the Graphs menu, define how the chart should appear, and then click OK. Here are the steps we followed to create the chart you see in Figure A.6.

1. Enter the data you want to use to create the chart.

2. Click Graphs → Legacy Dialogs → Bar. When you do this, you will see the Bar Charts dialog box you see in Figure A.15.

Figure A.15	The Bar Charts Dialog Box

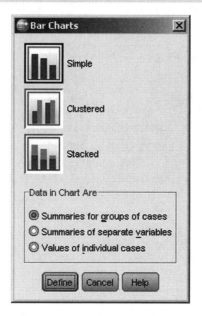

3. Click Simple.

4. Click Summaries for groups of cases.

5. Click Define. When you do this, you will see the Define Simple Bar: Summaries for Groups of Cases dialog box.

6. Click Cum n of cases.

7. Click gender, then move the variable to the Category Axis area by dragging.

8. Click OK, and you see the results of the chart in Figure A.16.

That's just the beginning of the chart, and to make any changes, you have to use the chart editor tools. Let's learn how to save the chart and then move to making changes.

Figure A.16   A Simple Bar Chart

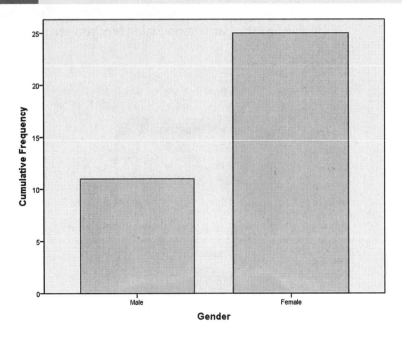

Saving a Chart

A chart is only one component of the Viewer window. A chart is part of the output generated when you perform some type of analysis. The chart is not a separate entity that stands by itself, and it cannot be saved as such. To save a chart, you need to save the contents of the entire Viewer. Follow these steps to do that:

1. Click File → Save.
2. Provide a name for the Viewer window.
3. Click OK. The output is saved under the name that you provide with a .spo extension.

ENHANCING SPSS CHARTS

Once you create a chart as we showed you in the previous section, you could finish the job by editing the chart to reflect exactly what you want to say. Colors, shapes, scales, fonts, and more can be changed. We will be working with the bar chart that was first shown to you in Figure A.6.

Editing a Chart

The first step in editing a chart is to double-click on the chart, then click the maximize button. You will see the entire chart in Figure A.17 in the Chart Editor window.

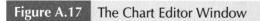
Figure A.17   The Chart Editor Window

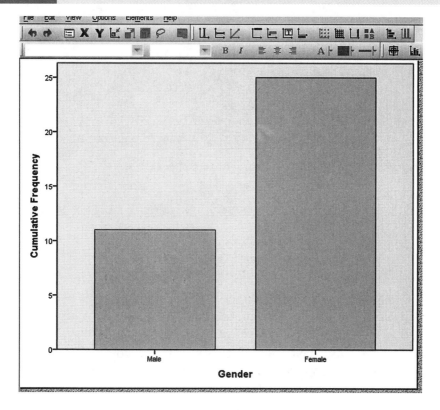

Working With Titles and Subtitles

Our first task is to enter a title and subtitle on the chart you saw in Figure A.16.

1. Click the Insert a Title icon on the Toolbar. When you do this, as you see in Figure A.18, you can then edit the element Title right on the screen and enter what you wish.

2. To insert a subtitle (or, in fact, as many titles as you like), just keep clicking the Insert a Title toolbar.

Figure A.18   Inserting a Title

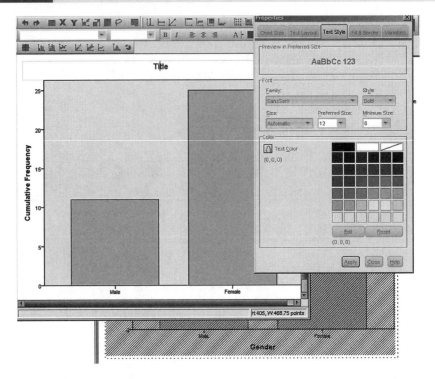

Working With Fonts

Once you have created a title or titles, you can work with fonts by double-clicking on the text you want to modify, and you will see the Properties dialog box as shown in Figure A.19. Click on the Text Style tab and you can make whatever changes you wish.

Working With Axes

The x- and y-axes provide the calibration for the independent (usually the x-axis) variable and the dependent (usually the y-axis) variable. SPSS names the y-axis the Scale axis and the x-axis the Category axis. Each of these axes can be modified in a variety of ways. To modify either axis, double-click on the title of the axis.

Figure A.19 | Working With Fonts

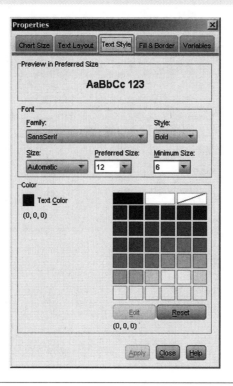

How to Modify the Scale (y) Axis

To modify the y-axis, follow these steps:

1. Still in the chart editor? We hope so. Double-click on the axis (not the axis label).

2. Click the Scale tab in the Properties dialog box. When you do this, you will see the Scale Axis dialog box, as shown in Figure A.20.

3. Select the options you want from the Scale Axis dialog box.

How to Modify the Category (x) Axis

Working with the x-axis is no more difficult than working with the y-axis. Here is how the x-axis was modified.

1. Double-click on the x-axis. The Category Axis dialog box opens. It is very similar to the Scale Axis dialog box that you saw in Figure A.20.

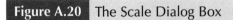

Figure A.20　The Scale Dialog Box

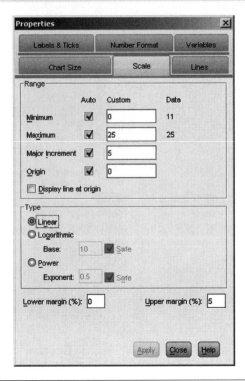

2. Select the options you want from the Category Axis dialog box.

When you are done, close the Chart Editor by double-clicking on the window icon or selecting File → Close.

DESCRIBING DATA

Now you have some idea about how data files are created in SPSS. Let's move on to some examples of simple analysis.

Frequencies and Crosstab Tables

Frequencies simply compute the number of times that a particular value occurs. Crosstabs allow you to compute the number of times that a value occurs when categorized by one or more dimensions, such as gender and age. Both frequencies and crosstabs are often reported first in research reports because they give the reader an overview of what the data look like. To compute frequencies, follow these steps. You should be in the Data Editor window.

1. Click Analyze → Descriptive Statistics → Frequencies. When you do this, you will see the Frequencies dialog box as shown in Figure A.21.

Figure A.21   The Frequencies Dialog Box

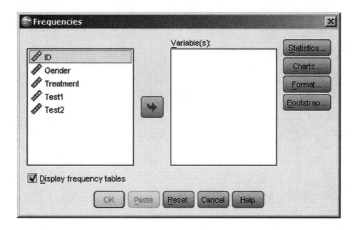

---

2. Double-click the variables for which you want frequencies computed. In this case, they are test1 and test2.

3. Click Statistics. You will see the Frequencies: Statistics dialog box, as shown in Figure A.22.

4. In the Dispersion area, click Std. deviation.

5. Under the Central Tendency area, click Mean.

6. Click Continue.

7. Click OK.

The output consists of a listing of the frequency of each value for test1 and test2, plus summary statistics (mean and standard deviation) for each, as you see in Figure A.23.

| Figure A.22 | The Frequencies: Statistics Dialog Box |

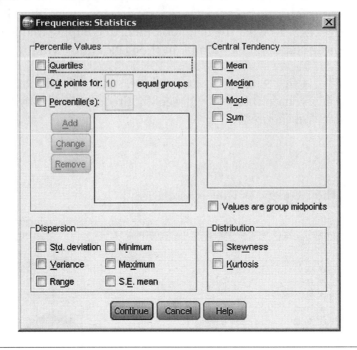

| Figure A.23 | Summary Statistics for test1 and test2 |

Statistics

		Test1	Test2
N	Valid	25	25
	Missing	0	0
Mean		83.52	64.24
Std. Deviation		8.627	21.642

Applying the Independent Samples t Test

Independent *t* tests are used to analyze data from a number of types of studies, including experimental, quasi-experimental, and field studies such as those shown in the following example, where we test the hypothesis that there are differences between males and females in reading.

How to Conduct an Independent Samples t Test

To conduct an independent *t* test, follow these steps.

1. Click Analyze → Compare Means → Independent-Samples T Test. When you do this, you will see the Independent-Samples T Test dialog box, as shown in Figure A.24.

Figure A.24	The Independent Samples T Test Frequencies Dialog Box

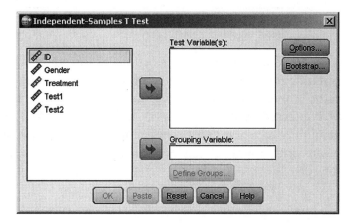

The Independent Samples t-Test Dialog Box

On the left-hand side of the dialog box, you see a listing of all the variables that can be used in the analysis. What you now need to do is define the test and the grouping variable.

2. Click test1, then drag it to the Test Variables area.

3. Click gender, then drag it to the Grouping Variable area.

4. Click Define Groups.

5. In the Group 1 box, type 1.

6. In the Group 2 box, type 2.

7. Click OK.

The output contains the means and standard deviations for each variable, plus the results of the *t* test, as shown in Figure A.25.

Figure A.25	Results of the Simple *t* Test

Group Statistics

	Gender	N	Mean	Std. Deviation	Std. Error Mean
Test1	Male	11	82.55	9.234	2.784
	Female	14	84.29	8.389	2.242

Independent Samples Test

		Levene's Test for Equality of Variances		t-test for Equality of Means							
										95% Confidence Interval of the Difference	
		F	Sig.	t	df	Sig. (2-tailed)	Mean Difference	Std. Error Difference	Lower	Upper
Test1	Equal variances assumed	.714	.407	-.493	23	.627	-1.740	3.532	-9.047	5.566
	Equal variances not assumed			-.487	20.532	.632	-1.740	3.575	-9.185	5.704

We have just given you the briefest of introductions to SPSS, and certainly none of these skills means anything if you don't know the value or meaningfulness of the data you originally entered. So, don't be impressed by your or others' skills at using programs like SPSS. Be impressed when those other people can tell you what the output means and how it reflects on your original question. And be really impressed if you can do it!

EXITING SPSS

To exit SPSS, click File → Exit. SPSS will be sure that you get the chance to save any unsaved or edited windows and will then close.
Ta-da! You're done!

Snapshots

"He wandered out of that building, fell to the ground, and has been spouting out random numbers ever since…"

APPENDIX B

Tables

TABLE B.1: AREAS BENEATH THE NORMAL CURVE

How to use this table:

1. Compute the z score based on the raw score and the mean of the sample.

2. Read to the right of the z score to determine the percentage of area underneath the normal curve or the area between the mean and computed z score.

Table B.1 Areas Beneath the Normal Curve

z-score	Area Between the Mean and the z-score	z-score	Area Between the Mean and the z-score	z-score	Area Between the Mean and the z-score	z-score	Area Between the Mean and the z-score	z-score	Area Between the Mean and the z-score	z-score	Area Between the Mean and the z-score	z-score	Area Between the Mean and the z-score	z-score	Area Between the Mean and the z-score
0.00	0.00	0.50	19.15	1.00	34.13	1.50	43.32	2.00	47.72	2.50	49.38	3.00	49.87	3.50	49.98
0.01	0.40	0.52	19.50	1.01	34.38	1.51	43.45	2.01	47.78	2.51	49.40	3.01	49.87	3.51	49.98
0.02	0.50	0.53	19.85	1.02	34.61	1.52	43.57	2.02	47.83	2.52	49.41	3.02	49.87	3.52	49.98
0.03	1.20	0.54	20.19	1.03	34.85	1.53	43.70	2.03	47.88	2.53	49.43	3.03	49.88	3.53	49.98
0.04	1.60	0.55	20.54	1.04	35.08	1.54	43.82	2.04	47.93	2.54	49.45	3.04	49.88	3.54	49.98
0.05	1.99	0.56	20.88	1.05	35.31	1.55	43.94	2.05	47.98	2.55	49.46	3.05	49.89	3.55	49.98
0.06	2.39	0.57	21.23	1.06	35.54	1.56	44.06	2.06	48.03	2.56	49.48	3.06	49.89	3.56	49.98
0.07	2.79	0.58	21.57	1.07	35.77	1.57	44.18	2.07	48.08	2.57	49.49	3.07	49.89	3.57	49.98
0.08	3.19	0.59	21.90	1.08	35.99	1.58	44.29	2.08	48.12	2.58	49.51	3.08	49.90	3.58	49.98
0.09	3.59	0.60	22.24	1.09	36.21	1.59	44.41	2.09	48.17	2.59	49.52	3.09	49.90	3.59	49.98
0.10	3.98	0.61	22.57	1.10	36.43	1.60	44.52	2.10	48.21	2.60	49.53	3.10	49.90	3.60	49.98
0.11	4.38	0.62	22.91	1.11	36.65	1.61	44.63	2.11	48.26	2.61	49.55	3.11	49.91	3.61	49.98
0.12	4.78	0.63	23.24	1.12	36.86	1.62	44.74	2.12	48.30	2.62	49.56	3.12	49.91	3.62	49.98
0.13	5.17	0.64	23.57	1.13	37.08	1.63	44.84	2.13	48.34	2.63	49.57	3.13	49.91	3.63	49.98
0.14	5.57	0.65	23.89	1.14	37.29	1.64	44.95	2.14	48.38	2.64	49.59	3.14	49.92	3.64	49.98
0.15	5.96	0.66	24.54	1.15	37.49	1.65	45.05	2.15	48.42	2.65	49.60	3.15	49.92	3.65	49.98
0.16	6.36	0.67	24.86	1.16	37.70	1.66	45.15	2.16	48.46	2.66	49.61	3.16	49.92	3.66	49.98
0.17	6.75	0.68	25.17	1.17	37.90	1.67	45.25	2.17	48.50	2.67	49.62	3.17	49.92	3.67	49.98
0.18	7.14	0.69	25.49	1.18	38.10	1.68	45.35	2.18	48.54	2.68	49.63	3.18	49.93	3.68	49.98
0.19	7.53	0.70	25.80	1.19	38.30	1.69	45.45	2.19	48.57	2.69	49.64	3.19	49.93	3.69	49.98
0.20	7.93	0.71	26.11	1.20	38.49	1.70	45.54	2.20	48.61	2.70	49.65	3.20	49.93	3.70	49.99
0.21	8.32	0.72	26.42	1.21	38.69	1.71	45.64	2.21	48.64	2.71	49.66	3.21	49.93	3.71	49.99
0.22	8.71	0.73	26.73	1.22	38.88	1.72	45.73	2.22	48.68	2.72	49.67	3.22	49.94	3.72	49.99
0.23	9.10	0.74	27.04	1.23	39.07	1.73	45.82	2.23	48.71	2.73	49.68	3.23	49.94	3.73	49.99

(Continued)

z-score	Area Between the Mean and the z-score	z-score	Area Between the Mean and the z-score	z-score	Area Between the Mean and the z-score	z-score	Area Between the Mean and the z-score	z-score	Area Between the Mean and the z-score	z-score	Area Between the Mean and the z-score	z-score	Area Between the Mean and the z-score	z-score	Area Between the Mean and the z-score
0.24	9.48	0.75	27.34	1.24	39.25	1.74	45.91	2.24	48.75	2.74	49.69	3.24	49.94	3.74	49.99
0.25	9.99	0.76	27.64	1.25	39.44	1.75	45.99	2.25	45.78	2.75	49.70	3.25	49.94	3.75	49.99
0.26	10.26	0.77	27.94	1.26	39.62	1.76	46.08	2.26	48.81	2.76	49.71	3.26	49.94	3.76	49.99
0.27	10.64	0.78	28.23	1.27	39.80	1.77	46.16	2.27	48.84	2.77	49.72	3.27	49.94	3.77	49.99
0.28	11.03	0.79	28.52	1.28	39.97	1.78	46.25	2.28	48.87	2.78	49.73	3.28	49.94	3.78	49.99
0.29	11.41	0.80	28.81	1.29	40.15	1.79	46.33	2.29	48.90	2.79	49.74	3.29	49.94	3.79	49.99
0.30	11.79	0.81	29.10	1.30	40.32	1.80	46.41	2.30	48.93	2.80	49.74	3.30	49.95	3.80	49.99
0.31	12.17	0.82	29.39	1.31	40.49	1.81	46.49	2.31	48.96	2.81	49.75	3.31	49.95	3.81	49.99
0.32	12.55	0.83	29.67	1.32	40.66	1.82	46.56	2.32	48.98	2.82	49.76	3.32	49.95	3.82	49.99
0.33	12.93	0.84	29.95	1.33	40.82	1.83	46.64	2.33	49.01	2.83	49.77	3.33	49.95	3.83	49.99
0.34	13.31	0.85	30.23	1.34	40.99	1.84	46.71	2.34	49.04	2.84	49.77	3.34	49.95	3.84	49.99
0.35	13.68	0.86	30.51	1.35	41.15	1.85	46.78	2.35	49.06	2.85	49.78	3.35	49.96	3.85	49.99
0.36	14.06	0.87	30.78	1.36	41.31	1.86	46.86	2.36	49.09	2.86	49.79	3.36	49.96	3.86	49.99
0.37	14.43	0.88	31.06	1.37	41.47	1.87	46.93	2.37	49.11	2.87	49.79	3.37	49.96	3.87	49.99
0.38	14.80	0.89	31.33	1.38	41.62	1.88	46.99	2.38	49.13	2.88	49.80	3.38	49.96	3.88	49.99
0.39	15.17	0.90	31.59	1.39	41.77	1.89	47.06	2.39	49.16	2.89	49.81	3.39	49.96	3.89	49.99
0.40	15.54	0.91	31.86	1.40	41.92	1.90	47.13	2.40	49.18	2.90	49.81	3.40	49.97	3.90	49.99
0.41	15.91	0.92	32.12	1.41	42.07	1.91	47.19	2.41	49.20	2.91	49.82	3.41	49.97	3.91	49.99
0.42	16.28	0.93	32.38	1.42	42.22	1.92	47.26	2.42	49.22	2.92	49.82	3.42	49.97	3.92	49.99
0.43	16.64	0.94	32.64	1.43	42.36	1.93	47.32	2.43	49.25	2.93	49.83	3.43	49.97	3.93	49.99
0.44	17.00	0.95	32.89	1.44	42.51	1.94	47.38	2.44	49.27	2.94	49.84	3.44	49.97	3.94	49.99
0.45	17.36	0.96	33.15	1.45	42.65	1.95	47.44	2.45	49.29	2.95	49.84	3.45	49.98	3.95	49.99
0.46	17.72	0.97	33.40	1.46	42.79	1.96	47.50	2.46	49.31	2.96	49.85	3.46	49.98	3.96	49.99
0.47	18.08	0.98	33.65	1.47	42.92	1.97	47.56	2.47	49.32	2.97	49.85	3.47	49.98	3.97	49.99
0.48	18.44	0.99	33.89	1.48	43.06	1.98	47.61	2.48	49.34	2.98	49.86	3.48	49.98	3.98	49.99
0.49	18.79	1.00	34.13	1.49	43.19	1.99	47.67	2.49	49.36	2.99	49.86	3.49	49.98	3.99	49.99

TABLE B.2: T VALUES NEEDED FOR REJECTION OF THE NULL HYPOTHESIS

How to use this table:

1. Compute the t value test statistic.

2. Compare the obtained t value to the critical value listed in this table. Be sure you have calculated the number of degrees of freedom correctly and you have selected an appropriate level of significance.

3. If the obtained value is greater than the critical or tabled value, the null hypothesis (that the means are equal) is not the most attractive explanation for any observed differences.

4. If the obtained value is less than the critical or table value, the null hypothesis is the most attractive explanation for any observed differences.

Table B.2 t Values Needed for Rejection of the Null Hypothesis

df	One-Tailed Test 0.10	0.05	0.01	df	One-Tailed Test 0.10	Two-Tailed Test 0.05	0.01	Two-Tailed Test 0.05	0.01
1	3.078	6.314	31.821	1	6.314	12.706	63.657		
2	1.886	2.92	6.965	2	2.92	4.303	9.925		
3	1.638	2.353	4.541	3	2.353	3.182	5.841		
4	1.533	2.132	3.747	4	2.132	2.776	4.604		
5	1.476	2.015	3.365	5	2.015	2.571	4.032		
6	1.44	1.943	3.143	6	1.943	2.447	3.708		
7	1.415	1.895	2.998	7	1.895	2.365	3.5		
8	1.397	1.86	2.897	8	1.86	2.306	3.356		
9	1.383	1.833	2.822	9	1.833	2.262	3.25		
10	1.372	1.813	2.764	10	1.813	2.228	3.17		
11	1.364	1.796	2.718	11	1.796	2.201	3.106		
12	1.356	1.783	2.681	12	1.783	2.179	3.055		
13	1.35	1.771	2.651	13	1.771	2.161	3.013		
14	1.345	1.762	2.625	14	1.762	2.145	2.977		
15	1.341	1.753	2.603	15	1.753	2.132	2.947		
16	1.337	1.746	2.584	16	1.746	2.12	2.921		
17	1.334	1.74	2.567	17	1.74	2.11	2.898		
18	1.331	1.734	2.553	18	1.734	2.101	2.879		
19	1.328	1.729	2.54	19	1.729	2.093	2.861		
20	1.326	1.725	2.528	20	1.725	2.086	2.846		
21	1.323	1.721	2.518	21	1.721	2.08	2.832		
22	1.321	1.717	2.509	22	1.717	2.074	2.819		

	One-Tailed Test				Two-Tailed Test		
df	0.10	0.05	0.01	df	0.10	0.05	0.01
23	1.32	1.714	2.5	23	1.714	2.069	2.808
24	1.318	1.711	2.492	24	1.711	2.064	2.797
25	1.317	1.708	2.485	25	1.708	2.06	2.788
26	1.315	1.706	2.479	26	1.706	2.056	2.779
27	1.314	1.704	2.473	27	1.704	2.052	2.771
28	1.313	1.701	2.467	28	1.701	2.049	2.764
29	1.312	1.699	2.462	29	1.699	2.045	2.757
30	1.311	1.698	2.458	30	1.698	2.043	2.75
35	1.306	1.69	2.438	35	1.69	2.03	2.724
40	1.303	1.684	2.424	40	1.684	2.021	2.705
45	1.301	1.68	2.412	45	1.68	2.014	2.69
50	1.299	1.676	2.404	50	1.676	2.009	2.678
55	1.297	1.673	2.396	55	1.673	2.004	2.668
60	1.296	1.671	2.39	60	1.671	2.001	2.661
65	1.295	1.669	2.385	65	1.669	1.997	2.654
70	1.294	1.667	2.381	70	1.667	1.995	2.648
75	1.293	1.666	2.377	75	1.666	1.992	2.643
80	1.292	1.664	2.374	80	1.664	1.99	2.639
85	1.292	1.663	2.371	85	1.663	1.989	2.635
90	1.291	1.662	2.369	90	1.662	1.987	2.632
95	1.291	1.661	2.366	95	1.661	1.986	2.629
100	1.29	1.66	2.364	100	1.66	1.984	2.626
Infinity	1.282	1.645	2.327	Infinity	1.645	1.96	2.576

TABLE B.3: CRITICAL VALUES FOR ANALYSIS OF VARIANCE OR F TEST

How to use this table:

1. Compute the F value.

2. Determine the number of degrees of freedom for the numerator $(k - 1)$ and the number of degrees of freedom for the denominator $(n - k)$.

3. Locate the critical value by reading across to locate the degrees of freedom in the numerator and down to locate the degrees of freedom in the denominator. The critical value is at the intersection of this column and row.

4. If the obtained value is greater than the critical or tabled value, the null hypothesis (that the means are equal to one another) is not the most attractive explanation for any observed differences.

5. If the obtained value is less than the critical or tabled value, the null hypothesis is the most attractive explanation for any observed differences.

| Table B.3 | Critical Values for Analysis of Variance or *F* Test |

df for the Denominator	Type I Error Rate	df for the Numerator					
		1	**2**	**3**	**4**	**5**	**6**
1	.01	4052.00	4999.00	5403.00	5625.00	5764.00	5859.00
	.05	162.00	200.00	216.00	225.00	230.00	234.00
	.10	39.90	49.50	53.60	55.80	57.20	58.20
2	.01	98.50	99.00	99.17	99.25	99.30	99.33
	.05	18.51	19.00	19.17	19.25	19.30	19.33
	.10	8.53	9.00	9.16	9.24	9.29	9.33
3	.01	34.12	30.82	29.46	28.71	28.24	27.91
	.05	10.13	9.55	9.28	9.12	9.01	8.94
	.10	5.54	5.46	5.39	5.34	5.31	5.28
4	.01	21.20	18.00	16.70	15.98	15.52	15.21
	.05	7.71	6.95	6.59	6.39	6.26	6.16
	.10	.55	4.33	4.19	4.11	4.05	4.01
5	.01	16.26	13.27	12.06	11.39	10.97	10.67
	.05	6.61	5.79	5.41	5.19	5.05	4.95
	.10	4.06	3.78	3.62	3.52	3.45	3.41
6	.01	13.75	10.93	9.78	9.15	8.75	8.47
	.05	5.99	5.14	4.76	4.53	4.39	4.28
	.10	3.78	3.46	3.29	3.18	3.11	3.06
7	.01	12.25	9.55	8.45	7.85	7.46	7.19
	.05	5.59	4.74	4.35	4.12	3.97	3.87
	.10	3.59	3.26	3.08	2.96	2.88	2.83
8	.01	11.26	8.65	7.59	7.01	6.63	6.37
	.05	5.32	4.46	4.07	3.84	3.69	3.58
	.10	3.46	3.11	2.92	2.81	2.73	2.67
9	.01	10.56	8.02	6.99	6.42	6.06	5.80
	.05	5.12	4.26	3.86	3.63	3.48	3.37
	.10	3.36	3.01	2.81	2.69	2.61	2.55
10	.01	10.05	7.56	6.55	6.00	5.64	5.39
	.05	4.97	4.10	3.71	3.48	3.33	3.22
	.10	3.29	2.93	2.73	2.61	2.52	2.46

(Continued)

Table B.3 (Continued)

df for the Denominator	Type I Error Rate	df for the Numerator					
		1	2	3	4	5	6
11	.01	9.65	7.21	6.22	5.67	5.32	5.07
	.05	4.85	3.98	3.59	3.36	3.20	3.10
	.10	3.23	2.86	2.66	2.54	2.45	2.39
12	.01	9.33	6.93	5.95	5.41	5.07	4.82
	.05	4.75	3.89	3.49	3.26	3.11	3.00
	.10	3.18	2.81	2.61	2.48	2.40	2.33
13	.01	9.07	6.70	5.74	5.21	4.86	4.62
	.05	4.67	3.81	3.41	3.18	3.03	2.92
	.10	3.14	2.76	2.56	2.43	2.35	2.28
14	.01	8.86	6.52	5.56	5.04	4.70	4.46
	.05	4.60	3.74	3.34	3.11	2.96	2.85
	.10	3.10	2.73	2.52	2.40	2.31	2.24
15	.01	8.68	6.36	5.42	4.89	4.56	4.32
	.05	4.54	3.68	3.29	3.06	2.90	2.79
	.10	3.07	2.70	2.49	2.36	2.27	2.21
16	.01	8.53	6.23	5.29	4.77	4.44	4.20
	.05	4.49	3.63	3.24	3.01	2.85	2.74
	.10	3.05	2.67	2.46	2.33	2.24	2.18
17	.01	8.40	6.11	5.19	4.67	4.34	4.10
	.05	4.45	3.59	3.20	2.97	2.81	2.70
	.10	3.03	2.65	2.44	2.31	2.22	2.15
18	.01	8.29	6.01	5.09	4.58	4.25	4.02
	.05	4.41	3.56	3.16	2.93	2.77	2.66
	.10	3.01	2.62	2.42	2.29	2.20	2.13
19	.01	8.19	5.93	5.01	4.50	4.17	3.94
	.05	4.38	3.52	3.13	2.90	2.74	2.63
	.10	2.99	2.61	2.40	2.27	2.18	2.11
20	.01	8.10	5.85	4.94	4.43	4.10	3.87
	.05	4.35	3.49	3.10	2.87	2.71	2.60
	.10	2.98	2.59	2.38	2.25	2.16	2.09

df for the Denominator	Type I Error Rate	df for the Numerator					
		1	2	3	4	5	6
21	.01	8.02	5.78	4.88	4.37	4.04	3.81
	.05	4.33	3.47	3.07	2.84	2.69	2.57
	.10	2.96	2.58	2.37	2.23	2.14	2.08
22	.01	7.95	5.72	4.82	4.31	3.99	3.76
	.05	4.30	3.44	3.05	2.82	2.66	2.55
	.10	2.95	2.56	2.35	2.22	2.13	2.06
23	.01	7.88	5.66	4.77	4.26	3.94	3.71
	.05	4.28	3.42	3.03	2.80	2.64	2.53
	.10	2.94	2.55	2.34	2.21	2.12	2.05
24	.01	7.82	5.61	4.72	4.22	3.90	3.67
	.05	4.26	3.40	3.01	2.78	2.62	2.51
	.10	2.93	2.54	2.33	2.20	2.10	2.04
25	.01	7.77	5.57	4.68	4.18	3.86	3.63
	.05	4.24	3.39	2.99	2.76	2.60	2.49
	.10	2.92	2.53	2.32	2.19	2.09	2.03
26	.01	7.72	5.53	4.64	4.14	3.82	3.59
	.05	4.23	3.37	2.98	2.74	2.59	2.48
	.10	2.91	2.52	2.31	2.18	2.08	2.01
27	.01	7.68	5.49	4.60	4.11	3.79	3.56
	.05	4.21	3.36	2.96	2.73	2.57	2.46
	.10	2.90	2.51	2.30	2.17	2.07	2.01
28	.01	7.64	5.45	4.57	4.08	3.75	3.53
	.05	4.20	3.34	2.95	2.72	2.56	2.45
	.10	2.89	2.50	2.29	2.16	2.07	2.00
29	.01	7.60	5.42	4.54	4.05	3.73	3.50
	.05	4.18	3.33	2.94	2.70	2.55	2.43
	.10	2.89	2.50	2.28	2.15	2.06	1.99
30	.01	7.56	5.39	4.51	4.02	3.70	3.47
	.05	4.17	3.32	2.92	2.69	2.53	2.42
	.10	2.88	2.49	2.28	2.14	2.05	1.98
35	.01	7.42	5.27	4.40	3.91	3.59	3.37
	.05	4.12	3.27	2.88	2.64	2.49	2.37
	.10	2.86	2.46	2.25	2.14	2.02	1.95
40	.01	7.32	5.18	4.31	3.91	3.51	3.29
	.05	4.09	3.23	2.84	2.64	2.45	2.34
	.10	2.84	2.44	2.23	2.11	2.00	1.93

(Continued)

Table B.3   (Continued)

df for the Denominator	Type I Error Rate	df for the Numerator					
		1	2	3	4	5	6
45	.01	7.23	5.11	4.25	3.83	3.46	3.23
	.05	4.06	3.21	2.81	2.61	2.42	2.31
	.10	2.82	2.43	2.21	2.09	1.98	1.91
50	.01	7.17	5.06	4.20	3.77	3.41	3.19
	.05	4.04	3.18	2.79	2.58	2.40	2.29
	.10	2.81	2.41	2.20	2.08	1.97	1.90
55	.01	7.12	5.01	4.16	3.72	3.37	3.15
	.05	4.02	3.17	2.77	2.56	2.38	2.27
	.10	2.80	2.40	2.19	2.06	1.96	1.89
60	.01	7.08	4.98	4.13	3.68	3.34	3.12
	.05	4.00	3.15	2.76	2.54	2.37	2.26
	.10	2.79	2.39	2.18	2.05	1.95	1.88
65	.01	7.04	4.95	4.10	3.65	3.31	3.09
	.05	3.99	3.14	2.75	2.53	2.36	2.24
	.10	2.79	2.39	2.17	2.04	1.94	1.87
70	.01	7.01	4.92	4.08	3.62	3.29	3.07
	.05	3.98	3.13	2.74	2.51	2.35	2.23
	.10	2.78	2.38	2.16	2.03	1.93	1.86
75	.01	6.99	4.90	4.06	3.60	3.27	3.05
	.05	3.97	3.12	2.73	2.50	2.34	2.22
	.10	2.77	2.38	2.16	2.03	1.93	1.86
80	.01	3.96	4.88	4.04	3.56	3.26	3.04
	.05	6.96	3.11	2.72	2.49	2.33	2.22
	.10	2.77	2.37	2.15	2.02	1.92	1.85
85	.01	6.94	4.86	4.02	3.55	3.24	3.02
	.05	3.95	3.10	2.71	2.48	2.32	2.21
	.10	2.77	2.37	2.15	2.01	1.92	1.85
90	.01	6.93	4.85	4.02	3.54	3.23	3.01
	.05	3.95	3.10	2.71	2.47	2.32	2.20
	.10	2.76	2.36	2.15	2.01	1.91	1.84
95	.01	6.91	4.84	4.00	3.52	3.22	3.00
	.05	3.94	3.09	2.70	2.47	2.31	2.20
	.10	2.76	2.36	2.14	2.01	1.91	1.84
100	.01	6.90	4.82	3.98	3.51	3.21	2.99
	.05	3.94	3.09	2.70	2.46	2.31	2.19
	.10	2.76	2.36	2.14	2.00	1.91	1.83
Infinity	.01	6.64	4.61	3.78	3.32	3.02	2.80
	.05	3.84	3.00	2.61	2.37	2.22	2.10
	.10	2.71	2.30	2.08	1.95	1.85	1.78

TABLE B.4: VALUES OF THE CORRELATION COEFFICIENT NEEDED FOR REJECTION OF THE NULL HYPOTHESIS

How to use this table:

1. Compute the value of the correlation coefficient.

2. Compare the value of the correlation coefficient with the critical value listed in this table.

3. If the obtained value is greater than the critical or tabled value, the null hypothesis (that the correlation coefficient is equal to 0) is not the most attractive explanation for any observed differences.

4. If the obtained value is less than the critical or tabled value, the null hypothesis is the most attractive explanation for any observed differences.

Table B.4	Values of the Correlation Coefficient Needed for Rejection of the Null Hypothesis

One-Tailed Test			Two-Tailed Test		
df	.05	.01	df	.05	.01
1	.9877	.9995	1	.9969	.9999
2	.9000	.9800	2	.9500	.9900
3	.8054	.9343	3	.8783	.9587
4	.7293	.8822	4	.8114	.9172
5	.6694	.832	5	.7545	.8745
6	.6215	.7887	6	.7067	.8343
7	.5822	.7498	7	.6664	.7977
8	.5494	.7155	8	.6319	.7646
9	.5214	.6851	9	.6021	.7348
10	.4973	.6581	10	.5760	.7079
11	.4762	.6339	11	.5529	.6835
12	.4575	.6120	12	.5324	.6614
13	.4409	.5923	13	.5139	.6411
14	.4259	.5742	14	.4973	.6226
15	.412	.5577	15	.4821	.6055
16	.4000	.5425	16	.4683	.5897
17	.3887	.5285	17	.4555	.5751
18	.3783	.5155	18	.4438	.5614
19	.3687	.5034	19	.4329	.5487
20	.3598	.4921	20	.4227	.5368
25	.3233	.4451	25	.3809	.4869
30	.2960	.4093	30	.3494	.4487
35	.2746	.3810	35	.3246	.4182
40	.2573	.3578	40	.3044	.3932
45	.2428	.3384	45	.2875	.3721
50	.2306	.3218	50	.2732	.3541
60	.2108	.2948	60	.2500	.3248
70	.1954	.2737	70	.2319	.3017
80	.1829	.2565	80	.2172	.2830
90	.1726	.2422	90	.2050	.2673
100	.1638	.2301	100	.1946	.2540

TABLE B.5: CRITICAL VALUES FOR THE CHI-SQUARE TEST

How to use this table:

1. Compute the χ^2 value.

2. Determine the number of degrees of freedom for the rows $(R - 1)$ and the number of degrees of freedom for the columns $(C - 1)$. If it's a one-dimension table, then you have only columns.

3. Locate the critical value by locating the degrees of freedom in the titled (df) column, and under the appropriate column for level of significance, read across.

4. If the obtained value is greater than the critical or tabled value, the null hypothesis (that the frequencies are equal to one another) is not the most attractive explanation for any observed differences.

5. If the obtained value is less than the critical or tabled value, the null hypothesis is the most attractive explanation for any observed differences.

Table B.5	Critical Values for the Chi-Square Test

	Level of Significance		
df	.10	.05	.01
1	2.71	3.84	6.64
2	4.00	5.99	9.21
3	6.25	7.82	11.34
4	7.78	9.49	13.28
5	9.24	11.07	15.09
6	10.64	12.59	16.81
7	12.02	14.07	18.48
8	13.36	15.51	20.09
9	14.68	16.92	21.67
10	16.99	18.31	23.21
11	17.28	19.68	24.72
12	18.65	21.03	26.22
13	19.81	22.36	27.69
14	21.06	23.68	29.14
15	22.31	25.00	30.58
16	23.54	26.30	32.00
17	24.77	27.60	33.41
18	25.99	28.87	34.80
19	27.20	30.14	36.19
20	28.41	31.41	37.57
21	29.62	32.67	38.93
22	30.81	33.92	40.29
23	32.01	35.17	41.64
24	33.20	36.42	42.98
25	34.38	37.65	44.81
26	35.56	38.88	45.64
27	36.74	40.11	46.96
28	37.92	41.34	48.28
29	39.09	42.56	49.59
30	40.26	43.77	50.89

Snapshots

"Statistics show that crime rates rise with the sale of ice cream. Now do your part and move along…"

APPENDIX C

Data Sets

These data files are referred to throughout *Statistics for People Who (Think They) Hate Statistics*. They are available here to be entered manually, or you can download them from one of two places.

The first is the website hosted by SAGE at http://www.sagepub.com/salkind4e.

The second is the author's website at http://onlinefilefolder.com. The username is *ancillaries* and the password is *files*. You will be able to download the files from that location.

At both sites, you can download the data files in either SPSS or Excel format. Note that values (such as 1 and 2) are included and not labels (such as male and female) for those values. For example, for Chapter 9 Data Set 2, gender is represented by 1 (male) or 2 (female). If you use SPSS, you can use the labels feature to assign labels to these values.

Chapter 2 Data Set 1

Prejudice	Prejudice	Prejudice	Prejudice
87	87	76	81
99	77	55	82
87	89	64	99
87	99	81	93
67	96	94	94

Chapter 2 Data Set 2

Score 1	Score 2	Score 3
3	34	154
7	54	167
5	17	132
4	26	145
5	34	154
6	25	145
7	14	113
8	24	156
6	25	154
5	23	123

Chapter 2 Data Set 3

Number of Beds	Infection Rate	Number of Beds	Infection Rate
234	1.7	342	5.3
214	2.4	276	5.6
165	3.1	187	1.2
436	5.6	512	3.3
432	4.9	553	4.1

Chapter 3 Data Set 1

Reaction Time	Reaction Time	Reaction Time	Reaction Time	Reaction Time
0.4	0.3	1.1	0.5	0.5
0.7	1.9	1.3	2.6	0.7
0.4	1.2	0.2	0.5	1.1
0.9	2.8	0.6	2.1	0.9
0.8	0.8	0.8	2.3	0.6
0.7	0.9	0.7	0.2	0.2

Chapter 3 Data Set 2

Math Score	Reading Score	Math Score	Reading Score
78	24	72	77
67	35	98	89
89	54	88	76
97	56	74	56
67	78	58	78
56	87	98	99
67	65	97	83
77	69	86	69
75	98	89	89
68	78	69	73
78	85	79	60
98	69	87	96
92	93	89	59
82	100	99	89
78	98	87	87

Chapter 3 Data Set 3

Height	Weight	Height	Weight
53	156	57	154
46	131	68	166
54	123	65	153
44	142	66	140
56	156	54	143
76	171	66	156
87	143	51	173
65	135	58	143
45	138	49	161
44	114	48	131

Chapter 4 Data Set 1

Comprehension Score	Comprehension Score	Comprehension Score	Comprehension Score
12	36	49	54
15	34	45	56
11	33	45	57
16	38	47	59
21	42	43	54
25	44	31	56
21	47	12	43
8	54	14	44
6	55	15	41
2	51	16	42
22	56	22	7
26	53	29	
27	57	29	

Chapter 4 Data Set 2

Monday	Tuesday	Wednesday	Thursday	Friday
12	17	10	15	20
9	11	10	4	0
6	8	9	5	10
4	0	5	4	9
9	7	8	5	11
10	5	4	4	15
13	12	7	3	10
22	16	18	15	20
1	3	6	4	2
5	8	4	6	7
7	0	3	8	2
10	4	1	8	12
4	5	8	6	9
15	12	10	9	11
3	6	4	7	10

Chapter 5 Data Set 1

Income	Education	Income	Education
$36,577	11	$64,543	12
$54,365	12	$43,433	14
$33,542	10	$34,644	12
$65,654	12	$33,213	10
$45,765	11	$55,654	15
$24,354	7	$76,545	14
$43,233	12	$21,324	11
$44,321	13	$17,645	12
$23,216	9	$23,432	11
$43,454	12	$44,543	15

Chapter 5 Data Set 2

Number Correct	Attitude	Number Correct	Attitude
17	94	14	85
13	73	16	66
12	59	16	79
15	80	18	77
16	93	19	91

Chapter 6 Data Set 1

Fall Results	Spring Results	Fall Results	Spring Results
21	7	3	30
38	13	16	26
15	35	34	43
34	45	50	20
5	19	14	22
32	47	14	25
24	34	3	50
3	1	4	17
17	12	42	32
32	41	28	46
33	3	40	10
15	20	40	48
21	39	12	11
8	46	5	23

Chapter 11 Data Set 1

Group	Memory Test	Group	Memory Test	Group	Memory Test
1	7	1	5	2	3
1	3	1	7	2	2
1	3	1	1	2	5
1	2	1	9	2	4

(Continued)

(Continued)

Group	Memory Test	Group	Memory Test	Group	Memory Test
1	3	1	2	2	4
1	8	1	5	2	6
1	8	1	2	2	7
1	5	1	12	2	7
1	8	1	15	2	5
1	5	1	4	2	6
1	5	2	5	2	4
1	4	2	4	2	3
1	6	2	4	2	2
1	10	2	5	2	7
1	10	2	5	2	6
1	5	2	7	2	2
1	1	2	8	2	8
1	1	2	8	2	9
1	4	2	9	2	7
1	3	2	8	2	6
1	1	2	8	2	8
1	1	2	8	2	9
1	4	2	9	2	7
1	3	2	8	2	6

Chapter 11 Data Set 2

Gender	Hand Up	Gender	Hand Up
1	9	1	8
2	3	2	7
2	5	1	9
2	1	2	9
1	8	1	8
1	4	2	7

Gender	Hand Up	Gender	Hand Up
2	2	2	3
2	6	2	7
1	9	2	6
1	3	1	10
2	4	1	7
1	8	1	6
2	3	1	12
1	10	2	8
2	6	2	8

Chapter 11 Data Set 3

Group	Attitude	Group	Attitude
1	6.50	1	4.23
2	7.90	1	6.95
2	4.30	2	6.74
2	6.80	1	5.96
1	9.90	2	5.25
1	6.80	2	2.36
1	4.80	1	9.25
2	6.50	1	6.36
2	3.30	1	8.99
1	4.00	1	5.58
2	13.17	2	4.25
1	5.26	1	6.60
2	9.25	2	1.00
1	8.00	1	5.00
2	1.25	2	3.50

Chapter 11 Data Set 4

Group 1	3
Group 1	4
Group 1	10
Group 2	14
Group 2	7
Group 2	8
Group 2	10
Group 2	15
Group 2	9
Group 2	19
Group 2	9
Group 2	17
Group 2	18
Group 2	19
Group 2	8
Group 2	7
Group 2	9
Group 2	14

Chapter 12 Data Set 1

Pretest	Posttest	Pretest	Posttest	Pretest	Posttest
3	7	6	8	9	4
5	8	7	8	8	4
4	6	8	7	7	5
6	7	7	9	7	6
5	8	6	10	6	9

Pretest	Posttest	Pretest	Posttest	Pretest	Posttest
5	9	7	9	7	8
4	6	8	9	8	12
5	6	8	8		
3	7	9	8		

Chapter 12 Data Set 2

Before	After	Before	After
20	23	23	22
6	8	33	35
12	11	44	41
34	35	65	56
55	57	43	34
43	76	53	51
54	54	22	21
24	26	34	31
33	35	32	33
21	26	44	38
34	29	17	15
33	31	28	27
54	56		

Chapter 12 Data Set 3

Before	After	Before	After
1.30	6.50	9.00	8.40
2.50	8.70	7.60	6.40
2.30	9.80	4.50	7.20
8.10	10.20	1.10	5.80
5.00	7.90	5.60	6.90
7.00	6.50	6.20	5.90
7.50	8.70	7.00	7.60
5.20	7.90	6.90	7.80
4.40	8.70	5.60	7.30
7.60	9.10	5.20	4.60

Chapter 13 Data Set 1

Group	Language Score	Group	Language Score
1	87	2	81
1	86	2	82
1	76	2	78
1	56	2	85
1	78	2	91
1	98	3	89
1	77	3	91
1	66	3	96
1	75	3	87
1	67	3	89
2	87	3	90
2	85	3	89
2	99	3	96
2	85	3	96
2	79	3	93

Chapter 13 Data Set 2

Practice	Time	Practice	Time
1	58.7	2	54.6
1	55.3	2	51.5
1	61.8	2	54.7
1	49.5	2	61.4
1	64.5	2	56.9
1	61.0	3	68.0
1	65.7	3	65.9
1	51.4	3	54.7
1	53.6	3	53.6
1	59.0	3	58.7
2	64.4	3	58.7
2	55.8	3	65.7
2	58.7	3	66.5
2	54.7	3	56.7
2	52.7	3	55.4
2	67.8	3	51.5
2	61.6	3	54.8
2	58.7	3	57.2

Chapter 13 Data Set 3

Group	Score	Group	Score	Group	Score
First	2	Second	4	Third	13
First	5	Second	5	Third	10
First	7	Second	3	Third	16
First	8	Second	6	Third	11
First	5	Second	16	Third	11
First	4	Second	11	Third	7
First	3	Second	9	Third	3
First	6	Second	8	Third	10
First	5	Second	12	Third	15
First	4	Third	7	Third	12

Chapter 14 Data Set 1

Treatment	Gender	Loss	Treatment	Gender	Loss
1	1	76	2	1	88
1	1	78	2	1	76
1	1	76	2	1	76
1	1	76	2	1	76
1	1	76	2	1	56
1	1	74	2	1	76
1	1	74	2	1	76
1	1	76	2	1	98
1	1	76	2	1	88
1	1	55	2	1	78
1	2	65	2	2	65
1	2	90	2	2	67
1	2	65	2	2	67
1	2	90	2	2	87

Treatment	Gender	Loss	Treatment	Gender	Loss
1	2	65	2	2	78
1	2	90	2	2	56
1	2	90	2	2	54
1	2	79	2	2	56
1	2	70	2	2	54
1	2	90	2	2	56

Chapter 14 Data Set 2

Severity	Treatment	Pain Score	Severity	Treatment	Pain Score
1	Drug #1	6	2	Drug #2	7
1	Drug #1	6	2	Drug #2	5
1	Drug #1	7	2	Drug #2	4
1	Drug #1	7	2	Drug #2	3
1	Drug #1	7	2	Drug #2	4
1	Drug #1	6	2	Drug #2	5
1	Drug #1	5	2	Drug #2	4
1	Drug #1	6	2	Drug #2	4
1	Drug #1	7	2	Drug #2	3
1	Drug #1	8	2	Drug #2	3
1	Drug #1	7	2	Drug #2	4
1	Drug #1	6	2	Drug #2	5
1	Drug #1	5	2	Drug #2	6
1	Drug #1	6	2	Drug #2	7
1	Drug #1	7	2	Drug #2	7
1	Drug #1	8	2	Drug #2	6

(Continued)

(Continued)

Severity	Treatment	Pain Score	Severity	Treatment	Pain Score
1	Drug #1	9	2	Drug #2	5
1	Drug #1	8	2	Drug #2	4
1	Drug #1	7	2	Drug #2	4
1	Drug #1	7	2	Drug #2	5
2	Drug #1	7	1	Placebo	2
2	Drug #1	8	1	Placebo	1
2	Drug #1	8	1	Placebo	3
2	Drug #1	9	1	Placebo	4
2	Drug #1	8	1	Placebo	5
2	Drug #1	7	1	Placebo	4
2	Drug #1	6	1	Placebo	3
2	Drug #1	6	1	Placebo	3
2	Drug #1	6	1	Placebo	3
2	Drug #1	7	1	Placebo	4
2	Drug #1	7	1	Placebo	5
2	Drug #1	6	1	Placebo	3
2	Drug #1	7	1	Placebo	1
2	Drug #1	8	1	Placebo	2
2	Drug #1	8	1	Placebo	4
2	Drug #1	8	1	Placebo	3
2	Drug #1	9	1	Placebo	5
2	Drug #1	0	1	Placebo	4
2	Drug #1	9	1	Placebo	2
2	Drug #1	8	1	Placebo	3
1	Drug #2	6	2	Placebo	4

Severity	Treatment	Pain Score	Severity	Treatment	Pain Score
1	Drug #2	5	2	Placebo	5
1	Drug #2	4	2	Placebo	6
1	Drug #2	5	2	Placebo	5
1	Drug #2	4	2	Placebo	4
1	Drug #2	3	2	Placebo	4
1	Drug #2	3	2	Placebo	6
1	Drug #2	3	2	Placebo	5
1	Drug #2	4	2	Placebo	4
1	Drug #2	5	2	Placebo	2
1	Drug #2	5	2	Placebo	1
1	Drug #2	5	2	Placebo	3
1	Drug #2	6	2	Placebo	2
1	Drug #2	6	2	Placebo	2
1	Drug #2	7	2	Placebo	3
1	Drug #2	6	2	Placebo	4
1	Drug #2	5	2	Placebo	3
1	Drug #2	7	2	Placebo	2
1	Drug #2	6	2	Placebo	2
1	Drug #2	8	2	Placebo	1

Chapter 14 Data Set 3

Gender	Caff_Consumption	Stress
1	5	1
1	6	3
2	7	3
1	7	2

(Continued)

(Continued)

Gender	Caff_Consumption	Stress
1	5	3
1	6	
1	8	2
1	8	2
2	9	1
2	8	1
2	9	1
2	7	2
2	4	1
2	3	1
1	0	1
2	4	2
1	5	1
2	6	2
1	2	2
1	4	3
1	5	3
2	5	3
1	4	2
1	3	2
1	7	3
2	8	2
1	9	2
1	11	1
1	2	2
1	3	1

Chapter 15 Data Set 1

Quality of Marriage	Quality Parent-Child	Quality of Marriage	Quality Parent-Child
1	58.7	2	54.6
1	55.3	2	51.5
1	61.8	2	54.7
1	49.5	2	61.4
1	64.5	2	56.9
1	61.0	3	68.0
1	65.7	3	65.9
1	51.4	3	54.7
1	53.6	3	53.6
1	59.0	3	58.7
2	64.4	3	58.7
2	55.8	3	65.7
2	58.7	3	66.5
2	54.7	3	56.7
2	52.7	3	55.4
2	67.8	3	51.5
2	61.6	3	54.8
2	58.7	3	57.2

Chapter 15 Data Set 2

Motivation	GPA	Motivation	GPA
1	3.4	6	2.6
6	3.4	7	2.5
2	2.5	7	2.8
7	3.1	2	1.8
5	2.8	9	3.7
4	2.6	8	3.1
3	2.1	8	2.5
1	1.6	7	2.4
8	3.1	6	2.1
6	2.6	9	4.0

(Continued)

(Continued)

Motivation	GPA	Motivation	GPA
5	3.2	7	3.9
6	3.1	8	3.1
5	3.2	7	3.3
5	2.7	8	3.0
6	2.8	9	2.0

Chapter 15 Data Set 3

Income	Level of Education	Income	Level of Education
$45,675	1	$74,776	3
$34,214	2	$89,689	3
$67,765	3	$96,768	2
$67,654	3	$97,356	3
$56,543	2	$38,564	2
$67,865	1	$67,375	3
$78,656	3	$78,854	3
$45,786	2	$78,854	3
$87,598	3	$42,757	1
$88,656	3	$78,854	3

Chapter 15 Data Set 4

Hours of Study	Grade
0	80
5	93
8	97
6	100
5	75
3	83
4	98
8	100
6	90
2	78

Chapter 16 Data Set 1

Training	Injuries	Training	Injuries
12	8	11	5
3	7	16	7
22	2	14	8
12	5	15	3
11	4	16	7
31	1	22	3
27	5	24	8
31	1	26	8
8	2	31	2
16	2	12	2
14	7	24	3
26	2	33	3
36	2	21	5
26	2	12	7
15	6	36	3

Chapter 16 Data Set 2

Time	Correct	Time	Correct
14.5	5	13.9	3
13.4	7	17.3	12
12.7	6	12.5	5
16.4	2	16.7	4
21	4	22.7	3

Chapter 17 Data Set 1

Voucher	Voucher	Voucher	Voucher	Voucher
1	1	2	3	3
1	1	2	3	3
1	1	2	3	3
1	1	2	3	3
1	1	3	3	3
1	2	3	3	3
1	2	3	3	3
1	2	3	3	3
1	2	3	3	3
1	2	3	3	3
1	2	3	3	3
1	2	3	3	3
1	2	3	3	3
1	2	3	3	3
1	2	3	3	3
1	2	3	3	3
1	2	3	3	3
1	2	3	3	3

Chapter 17 Data Set 2

Gender	Gender	Gender	Gender	Gender
1	1	1	2	2
1	1	1	2	2
1	1	1	2	2
1	1	1	2	2
1	1	1	2	2
1	1	2	2	2
1	1	2	2	2
1	1	2	2	2
1	1	2	2	2
1	1	2	2	2
1	1	2	2	2
1	1	2	2	2
1	1	2	2	2
1	1	2	2	2
1	1	2	2	2
1	1	2	2	2
1	1	2	2	2
1	1	2	2	2
1	1	2	2	2
1	1	2	2	2

Sample Data Set

Gender	Treatment	Test1	Test2
1	1	98	32
2	2	87	33
2	1	89	54
2	1	88	44
1	2	76	64
1	1	68	54
2	1	78	44
2	2	98	32
2	2	93	64
1	2	76	37
2	1	75	43
2	1	65	56
1	1	76	78
2	1	78	99
2	1	89	87
2	2	81	56
1	1	78	78
2	1	83	56
1	1	88	67
2	1	90	88
1	1	93	81
1	2	89	93
2	2	86	87
1	1	77	80
1	1	89	99

Snapshots

Appendicession, *n.* The irresistible urge
to keep reading once everyone else has
long since moved on.

APPENDIX D

Answers to Practice Questions

CHAPTER 2

1.

Mean = 28.375

Median = 25.5

Mode = 23

2.

	Score 1	Score 2	Score 3
Mean	5.6	27.6	144.3
Median	5.5	25.0	149.5
Mode	5	25, 34	154

3. Here's the SPSS output

Statistics

		Hosp Size	Infection Rate
N	Valid	10	10
	Missing	10	0
Mean		335.10	3.7200

4. Here's what your one paragraph might look like.

As usual, the Chicken Littles [the mode] led the way in sales. The total amount of food sold was $303, for an average of $2.55 for each special.

5. Nothing too dramatic here in terms of values that are really big or small or look weird (all reasons to use the median), so we'll just use the mean. You can see the mean for the three stores as shown in the last row. Seems like these might be the numbers you want to approximate for your store to be around the average of all the stores you manage.

Average	Store 1	Store 2	Store 3	Store 4 New Store	Mean
Sales (in thousands)	$323.6	$234.6	$308.3	?	$288.83
Number of items purchased	3,454	5,645	4,565	?	4,554.67
Number of visitors	4,534	6,765	6,654	?	5,984.33

6. You use the median when you have extreme scores, which would disproportionately bias the mean. One situation where the median is preferable to the mean is where income is reported. Because it varies so much, you want a measure of central tendency that is insensitive to extreme scores. Another example is where you have an extreme score or an outlier, such as the speed with which a group of adolescents can run 100 yards, where there are one or two exceptionally fast individuals.

7. You would use the median because it is insensitive to extreme scores.

8.

Before Removal of Highest Score	Mean	$ 83,111
	Median	$ 77,153
After Removal of Highest Score	Mean	$ 75,318
	Median	$ 76,564

The median is the best measure of central tendency and it is the one score that best represents the entire set of scores. Why? Because it is relatively unaffected by the huge $199,000 (somewhat) extreme data point, but as you can see in the above table, the mean is not (it shot up to more than $83,000).

CHAPTER 3

1. The range is the most convenient measure of dispersion because it requires only that you subtract one number (the lowest value) from another number (the highest value). It's imprecise because it does not take into account the values that fall between the highest and the lowest values in a distribution. Use the range when you want a very gross (and not very precise) estimate of the variability in a distribution.

2.

High Score	Low Score	Inclusive Range	Exclusive Range
7	6	2	1
89	45	45	44
34	17	18	17
15	2	14	13
1	1	1	0

3. For the most part, first-year students have stopped growing by that time, and the enormous variability that one sees in early childhood and adolescence has evened out. On a personality measure, however, those individual differences seem to be constant and are expressed similarly at any age.

4. As individuals score more similarly, they are closer to the mean, and the deviation about the mean is smaller. Hence, the standard deviation is smaller as well.

5. The range is 30.

	Unbiased	Biased
s	10.19	9.60

The unbiased sample standard deviation equals 10.19. The biased estimate equals 9.60. The difference is due to dividing by a sample size of 8 (for the unbiased estimate) as compared to a sample size of 9 (for the biased estimate). The unbiased estimate of the variance is 103.78, and the biased estimate is 92.25.

6.

Test 1		Test 2		Test 3	
Mean	49.30	Mean	50.10	Mean	50.50
Median	48.00	Median	51.50	Median	52.00
Mode	48.00	Mode	53.00	Mode	52.00
Range	10.00	Range	8.00	Range	9.00
Stan dev	3.37	Stan dev	32.25	Stan dev	2.92
Variance	11.34	Variance	10.54	Variance	8.50

Test 3 has the highest average score and Test 3 also the smallest amount of variability.

7. The standard deviation is 1.64 and the variance is 2.68.

8. The standard deviation is the square root of the variance (which is 25), making the standard deviation 5. You can't possibly know what the range is by knowing only the standard deviation or the variance. You can't even tell if the range is large or small because you don't know what's being measured and you don't know the scale of the measurement (tiny little bugs, or output from steam engines).

9.

a. Range = 8, Standard Deviation = 3.16, Variance = 10

b. Range = .6, Standard Deviation = .26, Variance = .07

c. Range = 5.8, Standard Deviation = 2.22, Variance = 4.92

10. Here's a chart that summarizes the results. Look familiar? It should—it's exactly what the SPSS output looks like.

Statistics

		Height	Weight
N	Valid	20	20
	Missing	0	0
Std. Deviation		11.436	15.652
Variance		130.779	244.997
Range		43	59

11. OK, here's how you answer it. Compute the standard deviation for any set of 10 or so numbers by hand using the formula for the standard deviation that has $n - 1$ in the denominator (the unbiased estimate). Then compare it, using the same numbers, to the SPSS output. As you will see, they are the same, indicating that SPSS produces an unbiased estimate. If you got this one correct, you're smart—go to the head of the class.

CHAPTER 4

1.

   a. Here's the frequency distribution.

Class Interval	Frequency
45–50	1
40–44	2
35–39	3
30–34	8
25–29	10
20–24	10
15–19	8
10–14	4
5–9	2
0–4	2

Here's what the histogram (created using SPSS) should look like.

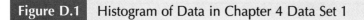

Figure D.1   Histogram of Data in Chapter 4 Data Set 1

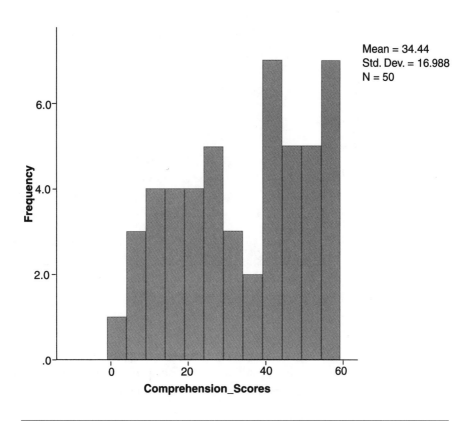

You may note that the *x*-axis in the histogram that you created is different from the one you see here. We double-clicked on it and, entering the Chart Editor, changed the number of ticks, range, and starting point for this axis. Nothing different—just looks a bit nicer.

b. We settled on a class interval of 5 because it allowed us to have close to 10 class intervals, and it fit the criteria that we discussed in this chapter for deciding on a class interval.

c. The distribution is negatively skewed because the mean is less than the median.

2. We used Excel and the Chart options to create a simple bar chart as you see in Figure D.2.

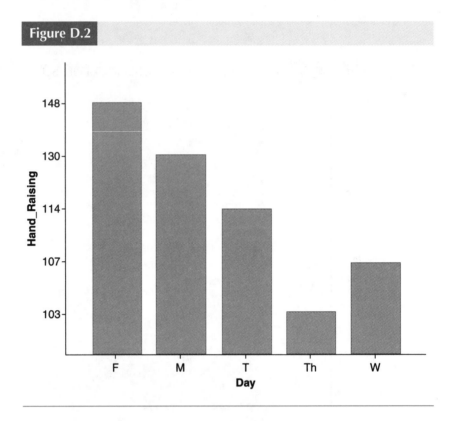

Figure D.2

3. The first thing you need to compute is the sum for each day (use the Analyze → Descriptive Statistics → Frequencies options) and then use those to create the bar graph.

4.

   a. This is negatively skewed because the majority of athletes scored in the upper range.

   b. Not skewed at all—in fact, the distribution is like a rectangle because everyone scored exactly the same.

   c. Positively skewed, because most of the spellers scored way low.

5.

   a. Pie

   b. Line

   c. Bar

   d. Line

   e. Bar

6. You will come up with examples of your own, but here are some of ours. Draw what you create.

   a. The number of words that a child knows as a function of ages from 12 months to 36 months.

   b. The percent of senior citizens who belong to the American Association for Retired Persons (AARP) as a function of gender and ethnicity.

   c. The proportion of students on scholarship at private and state-supported colleges.

7. On your own!

8. We did this using SPSS and the chart editor and it's as uninformative as it is ugly.

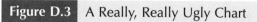

Figure D.3 A Really, Really Ugly Chart

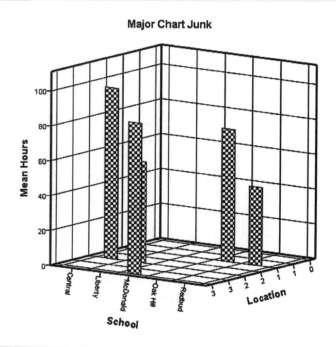

CHAPTER 5

1.

   a. $r = .596$.

   b. From the answer to 1a, you already know that the correlation is direct. But from the scatterplot shown in Figure D.4

Figure D.4   Scatterplot of Data in Data Set 2

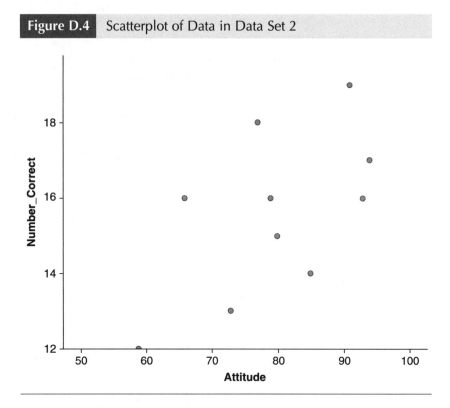

(we used SPSS, but you should do it by hand), you can predict it to be such (without actually knowing the sign of the coefficient) because the data points group themselves from the lower left corner of the graph to the upper right corner and assume a positive slope.

2.

a. $r = .269$.

b. According to the table presented earlier in the chapter, the general strength of the correlation of this magnitude is weak. The coefficient of determination is $.269^2$, so .072 or 7.2% of the variance is accounted for. The subjective analysis (weak) and the objective one (7.2% of the variance accounted for) are consistent with one another.

3. Note that the .47 has no sign, and in that case, we always assume that (like any other number) it is positive.

+.36

−.45

.47

−.62

+.71

4. The correlation is .64, meaning that increases in budget and increases in classroom achievement are positively related to one another (and note that we have to test for significance). And from a descriptive perspective, a bit more than 40% of the variance is shared between the two variables.

5. The correlation between number of minutes of exercise per day and GPA is .49, showing that as exercise increases, so does GPA, and of course, as exercise decreases, so does GPA.

6. The correlation is .14 and is so low because the set of GPA scores has very little variability. When there is so little variability, there is nothing to share and the two sets of scores have little in common—hence, the low correlation.

7.

   a. 8.

   b. Very strong.

   c. 1 – .64, or .36 (36%).

8. Here's the matrix . . .

	Age at Injury	Level of Treatment	12-Month Treatment Score
Age at Injury	1		
Level of Treatment	0.0557	1	
12-Month Treatment Score	–0.154	0.389	1

9. To examine the relationship between ethnicity and political affiliation, you would use the phi coefficient because both variables are nominal in nature. To examine the relationship between club membership and high school GPA, you would use the point biserial correlation because one variable is nominal (club membership) and the other is interval (GPA).

10. Just because two things are related does not mean that one causes the other. There are plenty of runners with average strength who can run fast, and plenty of very strong people who run slow. Strength can help make people run faster—but technique is more important (and, by the way, accounts for more of the variance).

CHAPTER 6

1. Do this one on your own.

2. Test-retest reliability should be established when you are interested in the consistency of an assessment over time, such as pre- and posttest types of studies or longitudinal studies. Parallel forms reliability is important to establish to make sure those different forms of the same test are similar to one another.

3. Test-retest reliability is established through the calculation of a simple correlation coefficient over the two testings. In this case, the correlation between fall and spring scores is .139, which has a probability of occurring by chance of .483. The correlation of .139 is not even close to what one might need (at least .85) for a conclusion that the test is reliable over time.

4. In general, a test that is reliable but not valid does what it does over and over, but does not do what it is supposed to. And, oops! A test cannot be valid without being reliable because if it does not do anything consistently, then it certainly cannot do one thing consistently.

5. It's simple. A test must first be able to do what it does over and over again (reliability) before one can conclude that it does what it should (validity). If a test is inconsistent (or unreliable), then it can't possibly be valid. For example, although a test of items like this one . . .

$$15 \times 3 = ?$$

would surely be reliable, if the 15 items on the test were to be labeled "Spelling Test," it surely would not be valid.

6. You need to use both a reliable and a valid test because if you get a null result, you will never be sure that the instrument is not measuring what it is supposed to rather than the hypothesis being faulty.

7. Content validity looks "on the face" of a test for whether it samples from an entire universe of possible items. For example, does a high school history test on the American Revolution contain items that reflect that subject area of American history?

Predictive validity examines how well a test predicts a particular outcome. For example, how well does a test of spatial skills predict success as a mechanical engineer?

Construct validity is present when a testing instrument assesses an underlying construct. For example, how well does an observation tool assess one dimension of manic-depression young adolescents?

Questions 1 and 4 are specific to your own interests. So, although there are no right answers, there are plenty of wrong ones!

CHAPTER 7

1. On your own!

2. The scientific method is kind of like a gift from the heavens. There's no one individual who can take credit for its discovery, but it probably came to be as we know it today from a variety of influences, ideas, and perspectives.

   Why does it work? It works because it allows the scientists (that's you) to rule out competing sources of explanation for some observed phenomenon. You can look at all the possible variables that may be important in any scientific endeavor and test them, ignore them, or control for their influence. When you do one or more of these steps, you can then find out if changes in *x* are indeed responsible for changes in *y*.

3. A good sample is highly representative of the population from which it is selected. When it is a good sample, it means that you can get a more true reading of actual population characteristics, and everything from your most basic findings to your inference to other populations is increased in accuracy.

   Poor sample selection = crummy population representation. Tattoo that on your forearm.

4.

   a. *Null:* Children with short attention spans, as measured by the Attention Span Observation Scale, will have the same frequency of out-of-seat behavior as those with long attention spans.

Directional: Children with short attention spans, as measured by the Attention Span Observation Scale, will have a higher frequency of out-of-seat behavior than those with long attention spans.

Nondirectional: Children with short attention spans, as measured by the Attention Span Observation Scale, differ in their frequency of out-of-seat behavior from those with long attention spans.

b. *Null:* There is no relationship between the overall quality of a marriage and spouses' relationships with their siblings.

Directional: There is a positive relationship between the overall quality of a marriage and spouses' relationships with their siblings.

Nondirectional: There is a relationship between the overall quality of a marriage and spouses' relationships with their siblings.

c. *Null:* Pharmacological treatment combined with traditional psychotherapy has the same effect in treating anorexia nervosa as does traditional psychotherapy alone.

Directional: Pharmacological treatment combined with traditional psychotherapy is more effective in treating anorexia nervosa than is traditional psychotherapy alone.

Nondirectional: Pharmacological treatment combined with traditional psychotherapy has a different effect from treating anorexia nervosa with traditional psychotherapy alone.

5.

a. Null hypothesis: There is no difference between the amount of money spent on food by undergraduate students and the amount spent by undergraduate student-athletes. $$H_0: \mu_{us} = \mu_{sa}$$	a. Research hypothesis: Undergraduate student-athletes spend more money on food than undergraduate students. $$H_1: X_{us} < X_{sa}$$
b. Null hypothesis: There is no difference between white and brown rats in the average amount of time taken to get out of a maze. $$H_0: \mu_W = \mu_B$$	b. Research hypothesis: There is a difference between white and brown rats in the average amount of time taken to get out of a maze. $$H_1: X_W \neq X_B$$

c. Null hypothesis: The effects of Drug A on a disease are not different from the effects of Drug B. H_0: $\mu_A = \mu_B$	c. Research hypothesis: Drug A has stronger effects on a disease than Drug B. H_1: $X_A > X_B$.
d. Null hypothesis: There is no difference between Method 1 and Method 2's time to complete a task. H_0: $\mu_1 = \mu_2$	d. Research hypothesis: There is a difference between Method 1 and Method 2's time to complete a task. H_1: $X_1 \neq X_2$

6. Well, if you're at the beginning of exploring a question (which then becomes a hypothesis) and you have little knowledge about the outcome (which is why you are asking the question and performing the test), then the null is the perfect starting point because it is a statement of equality that basically says, "Given no other information about the relationships that we are studying, I should start at the beginning, where I know very little." The null is the perfect, unbiased, and objective starting point because it is the place where everything is thought to be equal unless proven otherwise.

7. On your own!

8. As you already know, the null hypothesis states that there is no relationship between variables. Why? Simply, that's the best place to start given no other information. For example, if you are investigating the role of early events in the development of linguistic skills, then it is best to assume that they have no role whatsoever unless you prove otherwise. That's why we set out to *test* null hypotheses and not *prove* them. We want to be as unbiased as possible.

9. A research question answered using a one-tailed test might read as, "Does weight training increase swimming speed?" whereas a research question answered using a two-tailed test might read as, "Is there a difference between weight gain on two different diets?"

CHAPTER 8

1. In a normal curve, the mean, median, and mode are equal to one another; the curve is symmetrical about the mean; and the tails are asymptotic. Height and weight are examples, as are intelligence and problem-solving skills.

2. The raw score, the mean, and the standard deviation.

3. Because they all use the same metric—the standard deviation—and we can compare scores in units of standard deviations.

4. A z score is a standard score (and comparable to others of the same type of score) because it is based on the degree of variability within its respective distribution of scores. A z score is always a measure of the distance between the mean and some point on the x-axis (regardless of the mean and standard deviation differences from one distribution to the next), and because the same units are used (units of standard deviations), they can be compared to one another. That's the magic—comparability.

5.

Raw Score	z Score
68.0	−0.3
57.2	−1.6
82.0	1.5
84.1	1.8
69.0	−0.1
65.7	−0.5
85.0	1.9
83.4	1.7
72.0	0.3

6.

18	–1.01237
19	–0.88146
15	–1.40509
20	–0.75055
25	–0.096
31	0.68946
17	–1.14328
35	1.21309
27	0.16582
22	–0.48873
34	1.08218
29	0.42764
40	1.86764
33	0.95127
21	–0.61964

7.

a. The probability of a score falling between a raw score of 70 and a raw score of 80 is .5646. A z score for a raw score of 70 is –.78, and a z score for a raw score of 80 is .78. The area from the mean to a z score of .78 is 28.23%. The area between the two scores is 28.23 times 2, or 56.46%.

b. The probability of a score falling above a raw score of 80 is .2167. A z score for a raw score of 80 is .78. The area between the mean and a z score of .78 is 28.23%. The area below a z score of .78 is .50 + .2823, or .7823. The difference between 1 (the total area under the curve) and .7823 is .2177.

c. The probability of a score falling between a raw score of 81 and a raw score of 83 is .068. A z score for a raw score of 81 is .94, and a z score for a raw score of 83 is 1.25. The area from the mean to a z score of .94 is 32.64%. The area between the mean to a z score of 1.25 is 39.44%. The difference between the two is .3944 − .3264 = .068, or 6.8%.

d. The probability of a score falling below a raw score of 63 is .03. A z score for a raw score of 63 is −1.88. The area between the mean and a z score of −1.88 is 46.99%. The area below a z score of 1.88 is 1 − (.50 + .4699) = .03.

8. A little magic lets us solve for the raw score using the same formulas for computing the z score that you saw throughout this chapter. Here's the transformed formulas . . .

$$X = (s \times z) + \overline{X},$$

and taking this one step further, all we really need to know is the z score of 90% (or 40% in Table B1), which is 1.29.

So we have the following formula

$$X = (s \times z) + \overline{X}$$

or

$$X = 78 + (5.5 \times 1.29) = 85.095.$$

Jake is home free if he gets that score and, along with it, his certificate.

9. It doesn't because raw scores are not comparable to one another when they belong to different distributions. A raw score of 80 on the math test, where the class mean was 40, is just not comparable with an 80 on the essay writing skills test, where everyone got the one answer correct. Distributions, like people, are not always comparable to one another. Not everything (or everyone) is comparable to something else.

10. Here's the info with the unknown values in bold.

Noah has the higher average raw score (86.5 vs. 84 for Talya), but Talya has the higher average z score (1.2 vs. 1.05 for Noah). Remember that we asked who was the better student

Math			
Class Mean	81		
Class Standard Deviation	2		
Reading			
Class Mean	87		
Class Standard Deviation	10		
Raw Scores			
	Math Score	Reading Score	Average
Noah	85	88	86.5
Talya	87	81	84
z Scores			
	Math Score	Reading Score	Average
Noah	2	0.1	1.05
Talya	3	–0.6	1.2

relative to the rest, which requires the use of a standard score (we used z scores). But why is Talya the better student relative to Noah? It's because on the tests with the lowest variability (Math with an $sd = 2$), Talya really stands out with a z score of 3. That put her ahead to stay.

CHAPTER 9

1. The concept of significance is crucial to the study and use of inferential statistics because significance (reflected in the idea of a significance level) sets the level at which we can be confident that the outcomes we observe are

"truthful" and to what extent these outcomes can be generalized to the larger population from which the sample was selected.

2. The critical value represents the minimum value at which the null hypothesis is no longer the accepted explanation for any differences that are observed. It is the cut point, where obtained values that are more extreme indicate that there is no equality, but a difference (and the nature of that difference is dependent upon the questions being asked).

3.

a. Reject the null hypothesis. Because the level of significance is less than 5%, this means that there is a relationship between a person's choice of music and his crime rate.

b. Fail to reject the null hypothesis. The level of significance is greater than .05, meaning there is no relationship between the amount of coffee consumed and GPA.

c. Nope—no relationship with a probability that high.

4.

a. Level of significance refers only to a single, independent test of the null hypothesis and not to multiple tests.

b. It is impossible to set the error rate to 0 because it is not possible that we might not reject a null hypothesis when it is actually true. There's always that chance.

c. The level of risk that you are willing to take to reject the null hypothesis when it is true has nothing to do with the meaningfulness of the outcomes of your research. You can have a highly significant outcome that is meaningless, or have a relatively high Type I error rate (.10) and have a very meaningful finding.

5. At the .01 level, less room is left for errors or mistakes because the test is more rigorous. In other words, it is "harder" to find an outcome that is sufficiently removed from what you would expect by chance (the null hypothesis) when the probability associated with that income is smaller (such as .01) rather than larger (such as .05).

6. It's a fine point but one that people get pretty excited about. We can't "reject" the null because we never directly test it. Remember, nulls reflect population characteristics, and the

whole point is that we cannot directly test populations, only samples. If we can't test it, how can we reject it?

7. You can come up with the substance of these yourself, but how about the following examples?

a. A significant difference between two groups of readers was found, with the group that received intensive comprehension training outperforming the group that got no training (but the same amount of attention) on a test of reading comprehension.

b. Examining a huge sample (which is why the finding was significant), a researcher found a very strong positive correlation between shoe size and calories eaten on a daily basis. Silly, but true . . .

8. Chance is reflected in the degree of risk (Type I error) that we are willing to take in the possible rejection of a true null hypothesis. It is, first and foremost, always a plausible explanation for any difference that we might observe, and it is the most attractive explanation given no other information.

9.

a. The striped area represents values that are extreme enough that none of them reflects a finding that supports the null hypothesis.

b. A larger number of values reflecting a higher likelihood of a Type I error.

CHAPTER 10

1. The one-sample Z test is used when you want to compare a sample mean to a population parameter. Actually, think of it as a test to see if one number (which is a mean) belongs to a huge set of numbers.

2. The (BIG) Z is similar to the small z for one very good reason. It is a standard score. The z score has the sample standard deviation as the denominator, whereas the Z-test value has the standard error of the mean (or a measure of the variability of all the means from the population) as the denominator. In other words, they both use a standard measure that allows us

to use the normal curve table (in Appendix B) to understand how far the values are from what we would expect by chance alone.

3. For the following situations, write out in words a null hypothesis.

   a. Bob's weight loss on the chocolate-only diet is not representative of weight loss in a large population of middle-aged men who are on a protein-only diet.

   b. Rate of flu infections per thousand citizens for this past flu season is not comparable to the rate of infection over the past 50 seasons.

   c. Blair's costs for this month's expenses are different from his average costs over the course of a year.

4. The Z-test results in a value of 1.26—not nearly extreme enough (we would need a value of 1.96) to conclude that the average number of cases (15) is any different from that of the state (which is 16).

CHAPTER 11

1. The mean for boys equals 7.93, and the mean for girls equals 5.31. The obtained *t* value is 3.006, and the critical *t* value at the .05 level for rejection of the null hypothesis for a one-tailed test (boys *more* than girls, remember?) is 1.701. Conclusion? Boys raise their hands significantly more!

2. Now this is very interesting. We have the same exact data, of course, but a different hypothesis. Here, the hypothesis is that the number of times is *different* (not just more or less), necessitating a two-tailed test. So, using Table B2 and at the .01 level for a two-tailed test, the critical value is 2.764. The obtained value of 3.006 (same results as when you did the analysis for #1 above) does exceed what we would expect by chance, and given this hypothesis, there is a difference. So in comparison with one another, a one-tailed finding (see Question 1 above) need not be as extreme as a two-tailed finding given the same data to reach the same conclusion (that the research hypothesis is supported).

3.

a. $t(18) = 14.14$

b. $t(46) = 0.88$

c. $t(32) = 6.59$

4.

a. $t_{crit}(18) = 2.101$. Because the observed t value exceeds the critical t value, we reject the null hypothesis.

b. $t_{crit}(46) = 2.009$ (using the value for 50 degrees of freedom). Because the observed t value does not exceed the critical t value, we fail to reject the null hypothesis.

c. $t_{crit}(32) = 2.03$ (using the value for 35 degrees of freedom). Because the observed t value exceeds the critical t value, we reject the null hypothesis.

5. First, here's the output for the t test between independent samples that was conducted using SPSS.

Group Statistics

	Group	N	Mean	Std. Deviation	Std. Error Mean
Score	In Home	20	4.1500	2.20705	.49351
	Out of Home	20	5.5000	1.73205	.38730

	Levene's Test for Equality of Variances		t-test for Equality of Means						95% Confidence Interval of the Difference	
	F	Sig.	t	df	Sig (2-tailed)	Mean Difference	Std. Error Difference		Lower	Upper
Score Equal variances assumed	.938	.339	−2.152	38	.038	−1.35000	.62734		−2.61998	−.08002
Equal variances not assumed			−2.152	35.968	.038	−1.35000	.62734		−2.62234	−.07766

And here's one summary paragraph . . .

There was a significant difference between the mean performance for the group that received treatment in the home versus the group that received treatment out of the home. The mean for the In Home group was 4.15, and the mean for the Out of Home group was 5.5. The probability associated with this difference was .038, meaning that there is a less than 4% chance that the observed difference is due to chance. It is much more likely that the Out of Home program was more effective.

6. You can see the SPSS output in Figure D.5, where there is no significant difference ($p = .253$) between urban and rural dwellers in their attitude toward gun control.

Figure D.5   SPSS Output for a *t* Test of Independent Means

Group Statistics

	Group	N	Mean	Std. Deviation	Std. Error Mean
Attitude	Urban	16	6.5112	1.77221	.44305
	Rural	14	5.3979	3.31442	.88582

Independent Samples Test

		Levene's Test for Equality of Variances		t-test for Equality of Means					95% Confidence Interval of the Difference	
		F	Sig.	t	df	Sig. (2-tailed)	Mean Difference	Std. Error Difference	Lower	Upper
Attitude	Equal variances assumed	4.463	.044	1.168	28	.253	1.11339	.95311	-.83897	3.06576
	Equal variances not assumed			1.124	19.273	.275	1.11339	.99044	-.95763	3.18442

7. This certainly is a good one to think about, has many different "correct" answers, and raises a bunch of issues. If all you are concerned about is the level of Type I error, then we suppose that Dr. L's findings are more trustworthy because these results imply that one is less likely to make a Type I error. However, both of these findings are significant, even if one of them is marginally so, so if your personal system of evaluating these kinds of outcomes says, "I believe that significant is significant—that's what's important," then both should be considered equally valid and equally trustworthy. However, do remember to keep in mind that the meaningfulness of outcomes is critical as well (and you should get extra points if you bring this into the discussion). It seems to us that regardless of the level of Type I error, the results from both studies are highly meaningful because the program will probably result in safer children.

8. Here are the data plus the answers:

Experiment	Effect Size
1	2.6
2	1.3
3	0.65

As you can see, as the standard deviation doubles, the effect size is half as much. Why? If you remember, effect size gives you another indication of how meaningful the differences between groups are. If there is very little variability, there is not much difference between individuals, and any mean differences become more interesting (and probably more meaningful). When the standard deviation is 2 in our example, then the effect size is 2.6. But when the variability increases to 8 (the third experiment), interestingly, the effect size is reduced to .65. It's tough to talk about how meaningful differences between groups might be when there is less and less similarity between members of those groups.

9. And the answer? As you can see below, Group 2 scored a higher mean of 12.20 compared to a mean for Group 1 of 7.67), and based on the results you also see in Figure D.6, the difference is significant at the .004 level. Our conclusion is that the second group of fourth graders spells better.

Figure D.6

Group Statistics

	Group	N	Mean	Std. Deviation	Std. Error Mean
Score	Group 1	15	7.67	3.200	.826
	Group 2	15	12.20	4.539	1.172

Independent Samples Test

		Levene's Test for Equality of Variances		t-test for Equality of Means					95% Confidence Interval of the Difference	
		F	Sig.	t	df	Sig. (2-tailed)	Mean Difference	Std. Error Difference	Lower	Upper
Score	Equal variances assumed	5.482	.027	-3.162	28	.004	-4.533	1.434	-7.470	-1.596
	Equal variances not assumed			-3.162	25.159	.004	-4.533	1.434	-7.485	-1.581

CHAPTER 12

1. A *t* test for independent means tests two distinct groups of participants, and each group is tested once. A *t* test for dependent means tests one group of participants, and each participant is tested twice.

2.

   a. independent

   b. independent

   c. dependent

   d. independent

   e. dependent

3. The mean for before the recycling program was 34.44, and the mean for after was 34.84. There is an increase in recycling. Is the difference across the 25 districts significant? The obtained *t* value is .262, and with 24 degrees of freedom, the difference is not significant at the .01 level—the level at which the research hypothesis is being tested. Conclusion: The recycling program does not result in an increase in paper recycled.

4. Here's the SPSS output for this *t* test between dependent or paired means.

Paired Samples Statistics

	Mean	N	Std. Deviation	Std. Error Mean
Pair 1 Before_Treatment	32.8500	20	9.05117	2.02390
After_Treatment	36.9500	20	7.41602	1.65827

Paired Samples Test

	Paired Differences							
				95% Confidence Interval of the Difference				
	Mean	Std. Deviation	Std. Error Mean	Lower	Upper	t	df	Sig (2-tailed)
Pair 1 Before_Treatment After_Treatment	−4.10000	10.59245	2.36854	−9.05742	.85742	−1.731	19	.100

The conclusion? The before-treatment mean was lower (showing less tolerance) than the after-treatment mean. However, the difference between the two means is not significant, so the direction of the difference is irrelevant.

5. There was an increase in level of satisfaction, from 5.480 to 7.595, resulting in a t value of −3.893. This difference has an associated probability level of .001. It's very likely that the social service intervention worked.

6. The average score for Nibbles preference was 5.1, and the average score for Wribbles preference was 6.5. With 19 degrees of freedom, the t value for this test of dependent means was −1.965, with a critical value for rejecting the null of 2.093. Because the obtained t value of −1.965 does not exceed the critical value, the marketing consultant's conclusion is that both crackers are preferred about the same.

CHAPTER 13

1. Although both of these techniques look at differences between means, ANOVA is appropriate when more than two means are being compared. It can be used for a simple test between means, but it assumes that the groups are independent of one another.

2.

Design	Grouping Variable(s)	Test Variable
Simple ANOVA	Four levels of hours of training—2 , 4, 6, and 8 hours	Typing accuracy
	Three age groups—20-, 25- and 30-year-olds	Strength
	Six levels of job types	Job performance
Two-factor ANOVA	Two levels of training and gender (2×2 design)	Typing accuracy
	Three levels of age (5, 10, and 15 years) and number of siblings	Social skills
Three-factor ANOVA	Curriculum type (Type 1 or Type 2), GPA (above or below 3.0), and activity participation (participates or not)	ACT scores

3. The means for the three groups are 58.05 seconds, 57.96 seconds, and 59.03 seconds, and the probability of this F value $[F_{(2, 33)} = .160]$ occurring by chance is .853, far above what we would expect due to the treatment. Our conclusion? The number of hours of practice makes no difference in how fast you swim!

4. The F value is 7.013, which is significant at the .004 level, indicating that the amount of stress differs across the three groups.

CHAPTER 14

1. Easy. Factorial ANOVA is used only when you have more than one factor or independent variable! And actually, not so easy an answer to get (but if you get it, you really understand the material)—when you hypothesize an interaction.

2. Here's one of many different possible examples. There are three levels of one treatment (or factor) and two levels of severity of illness.

		Treatment		
		Drug #1	Drug #2	Placebo
Severity of Illness	Severe			
	Mild			

3. And the source table looks like this . . .

Tests of Between-Subjects Effects

Dependent Variable: PAIN_SCO

Source	Type III Sum of Squares	df	Mean Square	F	Sig.
Corrected Model	266.742	5	53.348	26.231	.000
Intercept	3070.408	1	3070.408	1509.711	.000
SEVERITY	.075	1	.075	.037	.848
TREATMEN	263.517	2	131.758	64.785	.000
SEVERITY * TREATMEN	3.150	2	1.575	.774	.463
Error	231.850	114	2.034		
Total	3569.000	120			
Corrected Total	498.592	119			

As far as our interpretation, in this data set, there is no main effect for severity, there is a main effect for treatment, and there is no interaction between the two main factors.

4.

   a. Yes. The high-stress group has an average of 38 cups consumed per month. The low- and no-stress groups both have an average of 30.5 cups consumed per month.

   b. No. The average for both the male and female groups is 33 cups consumed per month.

   c. There is a main effect for stress group, but not for gender. There is an interaction in these data.

CHAPTER 15

1.

   a. With 18 degrees of freedom ($df = n - 2$) at the .01 level, the critical value for rejection of the null hypothesis is .516. There is a significant correlation between speed and strength, and the correlation accounts for 32.15% of the variance.

   b. With 78 degrees of freedom at the .05 level, the critical value for rejection of the null hypothesis is .183 for a one-tailed test. There is a significant correlation between number correct and time. A one-tailed test was used because the research hypothesis was that the relationship was indirect or negative, and approximately 20% of the variance is accounted for.

   c. With 48 degrees of freedom at the .05 level, the critical value for rejection of the null hypothesis is .273 for a two-tailed test. There is a significant correlation between number of friends a child might have and GPA, and the correlation accounts for 13.69% of the variance.

2.

   a. and b. We used SPSS to compute the correlation as .434, significant at the .017 level using a two-tailed test. Figure D.7 shows the final output from the analysis.

Figure D.7	SPSS Output for Chapter 15 Data Set 2

Correlations

		Motivation	GPA
Motivation	Pearson Correlation	1	.434*
	Sig. (2-tailed)		.017
	N	30	30
GPA	Pearson Correlation	.434*	1
	Sig. (2-tailed)	.017	
	N	30	30

*Correlation is significant at the 0.05 level (2-tailed).

c. True. The more motivated you are, the more you will study; and the more you study, the more you are motivated. But (and this is a big "but") studying more does not cause you to be more highly motivated, nor does being more highly motivated cause you to study more.

3.

a. The correlation between income and education is .629, as you can see by examining the SPSS output shown here.

		Income	Level_Education
Income	Pearson Correlation	1	.629(**)
	Sig. (2-tailed)		.003
	N	20	20
Level_Education	Pearson Correlation	.629(**)	1
	Sig. (2-tailed)	.003	
	N	20	20

**Correlation is significant at the .01 level (2-tailed).

b. The correlation is significant at the .003 level.

c. The only argument you can make is that these two variables share something in common (and the more they share the higher the correlation) and that neither one can cause the other.

4.

a. $r_{(8)} = .69$.

b. $t_{crit}(8) = .6319$. The observed correlation of .69 exceeds the critical value using 8 degrees of freedom. Therefore, our observed correlation coefficient is statistically significant at the .05 level.

c. The shared variance equals 47% ($r^2_{hours.grade} = .47$).

d. There is a strong positive relationship between the number of hours spent studying and the grade earned on a test, which is statistically significant. The more hours a student studies, the higher his or her test grade; or the fewer hours spent studying, the lower the test grade!

5. The correlation is significant at the .01 level, and what's wrong with the statement is that an association between two variables does not imply that one causes the other. These two are correlated, which may be for many other reasons, but regardless of how much coffee one drinks, it does not result in a change in stress levels as a function of cause and effect.

6.
   a. The correlation is .763.

   b. With 8 degrees of freedom, and at the .05 level, the critical value for rejection of the null hypothesis that the correlation equals 0 is .5494 (see Table B.4). The obtained value of .763 is greater than the critical value (or what you would expect by chance), and our conclusion is that the correlation is significant and the two variables are related.

   c. If you recall, the best way to interpret any Pearson product-moment correlation is by squaring it, which gives us a coefficient of determination of .58, meaning that 58% of the variance in age is accounted for by the variance in number of words known. That's not huge, but as correlations between variables in human behavior go, it's pretty substantial.

7. The example here is the number of hours you study and your performance on your first test in statistics. These variables are not causally related. For example, you will have classmates who studied for hours and did poorly because they never understood the material, and classmates who did very well without any studying at all because they had some of the same material in another class. Just imagine if we forced someone to stay at his or her desk and study for 10 hours each of four nights before the exam. Would that ensure that he or she got a good grade? Of course not. Just because the variables are related does not mean that one causes the other.

CHAPTER 16

1. The major difference is that linear regression is used to explore if one variable predicts another. Analysis of variance examines differences between group mean, not prediction.

2.

    a. The regression equation is $Y' = -.214$ (number correct) + 17.202.

    b. $Y' = -.214(8) + 17.202 = 15.49$.

    c.

Time (Y)	# Correct (X)	Y′	Y – Y′
14.5	5	16.13	−1.6
13.4	7	15.70	−2.3
12.7	6	15.92	−3.2
16.4	2	16.77	−0.4
21.0	4	16.35	4.7
13.9	3	16.56	−2.7
17.3	12	14.63	2.7
12.5	5	16.13	−3.6
16.7	4	16.35	0.4
22.7	3	16.56	6.1

3.

    a. The other predictor variables should not be related to any other predictor variable. Only when they are independent of one another can they each contribute unique information to predicting the outcome or dependent variable.

    b. For example, living arrangements (single or in a group) and access to health care (high, medium, or low).

    c. Presence of Alzheimer's = (level of education)X_{IV1} + (general physical health)X_{IV2} + (living arrangements)X_{IV3} + (access to health care)X_{IV4} + a.

4. This one you do on your own.

5.

   a. You could compute the correlation between the two variables, which is .204. According to the information in Chapter 5, the magnitude of such a correlation is quite low. You could reach the conclusion that the number of wins is not a very good predictor of whether a team ever won a Super Bowl.

   b. Many variables are categorical by nature (gender, race, social class, and political party) and cannot be easily measured on a scale from 1 to 100, for example. Using categorical variables allows us more flexibility.

   c. Some other variables might be number of All-American players, win-loss record of coaches, and home attendance.

6.

   a. Caffeine consumption

   b. Stress group

   c. The correlation coefficient calculated from Chapter 15 is .373, which you would square to get .139, or R^2. Pretty nifty!

7.

   a. The best predictor of the three is years of experience, but because none is significant (.102), one could say that the three are equally good (or bad)!

   b. And here's the regression equation . . .

$$Y' = .959(X_1) - 5.786(X_2) - 1.839(X_3) + 96.337,$$

which, if we substitute the values for X_1, X_2, and X_3, we get the following . . .

$$Y' = .959(12) - 5.786(2) - 1.839(5) + 96.377$$

with a predicted chef's score of 87.12.

	Unstandardized Coefficients		Standardized Coefficients		
	B	Std. Error	Beta	t	Sig.
(Constant)	97.237	7.293		13.334	.000
Years_Ex	1.662	.566	1.163	2.937	.102
Level_Ed	−7.017	2.521	−.722	−2.783	.162
Num_Pos	−2.555	1.263	−.679	−2.022	.212

Note also that years of experience and level of education were significant predictors.

CHAPTER 17

1. When the expected and the observed values are identical. One example would be when you expected an unequal number of first-year and second-year students to show up and they really did!

2. Here's the worksheet for computing the chi-square value:

Category	O (observed frequency)	E (expected frequency)	D (difference)	$(O - E)^2$	$(O - E)^2/E$
Republican	800	800	0	0	0.00
Democrat	700	800	100	10,000	12.50
Independent	900	800	100	10,000 ·	12.50

With 2 degrees of freedom at the .05 level, the critical value needed for rejection of the null hypothesis is 5.99. The obtained value of 25 allows us to reject the null and conclude that there is a significant difference in the numbers of people who voted as a function of political party.

3. Here's the worksheet for computing the chi-square value:

Category	O (observed frequency)	E (expected frequency)	D (difference)	$(O - E)^2$	$(O - E)^2/E$
Boys	45	50	5	25	0.50
Girls	55	50	5	25	0.50

With 1 degree of freedom at the .01 level of significance, the critical value needed for rejection of the null hypothesis is 6.64. The obtained value of 1.00 means that the null cannot be rejected, and there is no difference between the number of boys and girls who play soccer.

4. Some fun facts first. The total number of students enrolled in all six grades is 2,217, and the expected frequency value of each cell is 2217/6 or 369.50.

The obtained chi-square value is 36.98. With 5 degrees of freedom at the .05 level, the value needed for rejection of the null hypothesis is 11.07. Because the obtained value of 36.98 exceeds the critical value, the conclusion is that the enrollment numbers are not what was expected, and indeed, there are significantly different proportions enrolled in each grade.

Snapshots

"As you can see by this pie chart, most
of our expenses go to, well, *pie*."

APPENDIX E

Math: Just the Basics

If you're reading this, then you know you may need a bit of help with your basic math skills. Lots of people need such help, especially after coming back to school after a break. There's nothing wrong with taking this little side trip before you continue working in *Statistics for People Who (Think They) Hate Statistics*.

Most of the skills you need to work the examples in this book and to complete those at the end of the chapters you already know. For example, you can add, subtract, multiply, and divide. You also probably know how to use a calculator to compute the square root of a number.

The confusion begins when we start dealing with equations and the various operations that can take place within parentheses, like these → ( ) and brackets, like these → [ ].

That's where we will spend most of our time and show you examples so you better understand how to work through what appear to be complex operations, but once they are reduced to their individual parts, completing them is a cinch.

THE BIG RULES SAY HELLO TO BODMAS

Sounds like an alien life form or something out of the Borg, right?

Nope. It is simply an acronym that indicates the order in which operations take place in an expression or an equation.

BODMAS goes like this . . .

B is for brackets, such as [ ], or sometimes parens (short for parentheses), like this ( ), both of which are sometimes present in an expression or equation.

O is for order or power, as in raise 4 to the order of 2 or 4^2.

D is division, as in 6/3

M is multiplication, as in 6×3

A is addition, as in 2 + 3

S is subtraction, as in 5 − 1

You perform these operations in the above order. For example, the first thing you do is work within brackets (or parentheses), then take numbers to a power (such as squaring), then divide, then multiply, and so on. If there's nothing to square, skip the *O* step, and if there's nothing to add, skip the *A* step. If both brackets and parens are present, do the brackets first, and then the parentheses.

For example, take a look at this simple expression . . .

$$(3 + 2) \times 2 = ?$$

Using the BODMAS acronym, we know that

1. the first thing we do is any operation within brackets. So, 3 + 2 = 5.

2. Moving on, the next available step is multiplying 5 times 2 for a total of 10, and 10 is the answer.

Here's another . . .

$$(4/2 \times 5) + 7 = ?$$

And here's what we do . . .

1. First is the division of 4 by 2, which equals 2, and that 2 is multiplied by 5 for a value of 10.

2. Then, 10 is added to 7 for a final sum of 17.

Let's get a bit fancier and use some squaring of numbers.

$$(10^2 \times 3)/150 = ?$$

Here's what we do . . .

1. Within brackets (as always for the first step).

2. Square 10 to equal 100, and then multiply it by 3 for a value of 300.

3. Divide 300 by 150 for a value of 2.

Let's add another layer of complexity when we have more than one set of brackets or parens. All you need remember is that we always begin inside the expression, solve the values within the brackets first as you work through the acronym.

Here we go . . .

$$[(15 \times 2) - (5 + 7)]/6 = ?$$

1. 15 times 2 is 30.

2. 5 plus 7 is 12.

3. 30 – 12 is 18.

4. 18 divided by 6 is 3.

THE LITTLE RULES

Just a few more. There are negative and positive numbers used throughout the book, and you need to know how they act when combined in different ways.

When multiplying negative and positive numbers, the following is true . . .

1. A negative number multiplied by a positive number (or a positive number multiplied by a negative number) always, always, always equals a negative number. For example...

$$-3 \times 2 = -6$$

or

$$4 \times -5 = -20$$

2. A negative number multiplied by another negative number results in a positive number. For example . . .

$$-4 \times -3 = 12$$

or

$$-2.5 \times -3 = 7.5$$

And, a negative number divided by a positive number results in a negative number. For example . . .

$$-10/2 = -5$$

or

$$25/-5 = 5$$

And finally, a negative number divided by a negative number results in a positive number. For example . . .

$$-10/-5 = 2$$

Practice makes perfect, so here are 10 problems with the answers following. If you can't get the right answer, review the order of operations above and also ask someone in your study group to review where you might have gone wrong.

Here we go . . .

1. $75 + 10 =$
2. $104 - 50 =$
3. $50 - 104 =$
4. 42×-2
5. -50×-60
6. $25/5 - 6/3$
7. $6,392 - (-700)$
8. $(510 - 500)/-10$
9. $[(40^2 - 207) - (80^2 - 400)]/35 \times 24$
10. $([(502 - 300) - 25] - [(242 - 100) - 50])/20 \times 30$

Answers

1. 85
2. 54
3. -54
4. -84
5. 3,000

6. 3

7. 7,092

8. −1

9. −5.48

10. .141

Want some more help and more practice? Take a look at these sites . . .

http://www.webmath.com/index.html

http://www.math.com/homeworkhelp/BasicMath.html

There's nothing worse than starting a course and being so anxious that any meaningful learning just can't take place. Thousands of students less well prepared than you have succeeded, and you can as well. Reread the Chapter 1 tips on how to approach the material in this course and good luck.

GLOSSARY

Analysis of variance

A test for the difference between two or more means. A simple analysis of variance (or ANOVA) has only one independent variable, whereas a factorial analysis of variance tests the means of more than one independent variable. One-way analysis of variance looks for differences between the means of more than two groups.

Arithmetic mean

A measure of central tendency that sums all the scores in the data sets and divides by the number of scores.

Asymptotic

The quality of the normal curve such that the tails never touch the horizontal axis.

Average

The most representative score in a set of scores.

Bell-shaped curve

A distribution of scores that is symmetrical about the mean, median, and mode and has asymptotic tails.

Class interval

The upper and lower boundary of a set of scores used in the creation of a frequency distribution.

Coefficient of alienation

The amount of variance unaccounted for in the relationship between two variables.

Coefficient of determination

The amount of variance accounted for in the relationship between two variables.

Coefficient of nondetermination

See Coefficient of alienation

Concurrent criterion validity

A type of validity that examines how well a test outcome is consistent with a criterion that occurs in the present.

Confidence interval

The best estimate of the range of a population value given the sample value.

Construct validity

A type of validity that examines how well a test reflects an underlying construct.

Content validity

A type of validity that examines how well a test samples a universe of items.

Correlation coefficient

A numerical index that reflects the relationship between two variables.

Correlation matrix

A set of correlation coefficients.

Criterion

Another term for the outcome variable.

Criterion validity

A type of validity that examines how well a test reflects some criterion that occurs in either the present (concurrent) or the future (predictive).

Critical value

The value necessary for rejection (or nonacceptance) of the null hypothesis.

Cumulative frequency distribution

A frequency distribution that shows frequencies for class intervals along with the cumulative frequency for each.

Data

A record of an observation or an event such as a test score, a grade in math class, or response time.

Data point

An observation.

Data set

A set of data points.

Degrees of freedom

A value that is different for different statistical tests and approximates the sample size of number of individual cells in an experimental design.

Dependent variable

The outcome variable or the predicted variable in a regression equation.

Descriptive statistics

Values that describe the characteristics of a sample or population.

Direct correlation

A positive correlation where the values of both variables change in the same direction.

Directional research hypothesis

A research hypothesis that includes a statement of inequality.

Error in prediction

The difference between the actual score (Y) and the predicted score (Y)

Error of estimate

See Error in prediction

Error score

The part of a test score that is random and contributes to the unreliability of a test.

Factorial analysis of variance

An analysis of variance with more than one factor or independent variable.

Factorial design

A research design where there is more than one treatment variable.

Frequency distribution

A method for illustrating the distribution of scores within class intervals.

Frequency polygon

A graphical representation of a frequency distribution.

Histogram

A graphical representation of a frequency distribution.

Hypothesis

An if-then statement of conjecture that relates variables to one another.

Independent variable

The treatment variable that is manipulated or the predictor variable in a regression equation.

Indirect correlation

A negative correlation where the values of variables move in opposite directions.

Inferential statistics

Tools that are used to infer the results based on a sample to a population.

Interaction effect

The outcome where the effect of one factor is differentiated across another factor.

Internal consistency reliability

A type of reliability that examines the one-dimensional nature of an assessment tool.

Interrater reliability

A type of reliability that examines the consistency of raters.

Interval level of measurement

A level of measurement that is defined by the equal appearance of spacing or values between points along the scale.

Kurtosis

The quality of a distribution such that it is flat or peaked.

Leptokurtic

The quality of a normal curve that defines its peakedness.

Line of best fit

The regression line that best fits the actual scores and minimizes the error in prediction.

Linear correlation

A correlation that is best expressed as a straight line.

Main effect

In analysis of variance, when a factor or an independent variable has a significant effect upon the outcome variable.

Mean

A type of average where scores are summed and divided by the number of observations.

Mean deviation

The average deviation for all scores from the mean of a distribution.

Measures of central tendency

The mean, median, and mode.

Median

The point at which 50% of the cases in a distribution fall below and 50% fall above.

Midpoint

The central point in a class interval.

Mode

The most frequently occurring score in a distribution.

Multiple regression

A statistical technique where several variables are used to predict one.

Nominal level of measurement

The most gross level of measurement where variables can be placed in categories.

Nondirectional research hypothesis

A hypothesis that posits no direction, but a difference.

Nonparametric statistics

Distribution-free statistics.

Normal curve

See Bell-shaped curve

Null hypothesis

A statement of equality between sets of variables.

Observed score

The score that is recorded or observed.

Obtained value

The value that results from the application of a statistical test.

Ogive

A visual representation of a cumulative frequency distribution.

One-sample *Z* test

Used for comparing a sample mean to a population mean.

One-tailed test

A directional test.

One-way analysis of variance

See Analysis of variance

Ordinal level of measurement

A level of measurement that is characterized by things being ordered.

Outliers

Those scores in a distribution that are noticeably much more extreme than the majority of scores. Exactly what score is an outlier is usually an arbitrary decision made by the researcher.

Parallel forms reliability

A type of reliability that examines the consistency across different forms of the same test.

Parametric statistics

Statistics used for the inference from a sample to a population.

Pearson product-moment correlation

See Correlation coefficient

Percentile point

The point at or below where a score appears.

Platykurtic

The quality of a normal curve that defines its flatness.

Population

All the possible subjects or cases of interest.

Post hoc

After the fact, referring to tests done to determine the true source of a difference between three or more groups.

Predictive validity

A type of validity that examines how well a test outcome is consistent with a criterion that occurs in the future.

Predictor

The variable that predicts an outcome.

Range

The highest minus the lowest score and a gross measure of variability. Exclusive range is the highest score minus the lowest score. Inclusive range is the highest score minus the lowest score plus 1.

Ratio level of measurement

A level of measurement defined as having an absolute zero.

Regression equation

The equation that defines the points and the line that are closest to the actual scores.

Regression line

The line drawn based on the values in the regression equation.

Reliability

The quality of a test such that it is consistent.

Research hypothesis

A statement of inequality between two variables.

Sample

A subset of a population.

Sampling error

The difference between sample and population values.

Scales of measurement

Different ways of categorizing measurement outcomes.

Scattergram or scatterplot

A plot of paired data points.

Significance level

The risk set by the researcher for rejecting a null hypothesis when it is true.

Simple analysis of variance

See Analysis of variance

Skew or skewness

The quality of a distribution that defines the disproportionate frequency of certain scores. A longer right tail than left corresponds to a smaller number of occurrences at the high end of the distribution; this is a positively skewed distribution. A shorter right tail than left corresponds to a larger number of occurrences at the high end of the distribution; this is a negatively skewed distribution.

Source table

A listing of sources of variance in an analysis of variance summary table.

Standard deviation

The average deviation from the mean.

Standard error of estimate

A measure of accuracy in prediction.

Standard score

See z score

Statistical significance

See Significance level

Statistics

A set of tools and techniques used to organize and interpret information.

Test-retest reliability

A type of reliability that examines consistency over time.

Test statistic value

See Obtained value

True score

The unobservable part of an observed score that reflects the actual ability or behavior.

Two-tailed test

A test of a nondirectional hypothesis where the direction of the difference is of little importance.

Type I error

The probability of rejecting a null hypothesis when it is true.

Type II error

The probability of accepting a null hypothesis when it is false.

Unbiased estimate

A conservative estimate of a population parameter.

Validity

The quality of a test such that it measures what it says it does.

Variability

The amount of spread or dispersion in a set of scores.

Variance

The square of the standard deviation, and another measure of a distribution's spread or dispersion.

Y' or Y prime

The predicted Y value.

z score

A raw score that is adjusted for the mean and standard deviation of the distribution from which the raw score comes.

INDEX